About Island Press

Wolves

and

Human Communities

Wolves
and
Human
Communities
Biology, Politics, and Ethics

Edited by
Virginia A. Sharpe, Bryan G. Norton, and Strachan Donnelley

ISLAND PRESS
Washington, D.C. • Covelo, California

Library of Congress Cataloging-in-Publication Data

Wolves and human communities : biology, politics, and ethics / [edited
by] Virginia A. Sharpe, Bryan Norton, Strachan Donnelley.
     p. cm.
Includes bibliographical references (p.   ).
   ISBN 1-55963-828-1 (cloth : alk. paper) — ISBN 1-55963-829-X (paper :
alk. paper)
   1. Wolves—New York—Adirondack Mountains. 2.
Wolves—Reintroduction—New York—Adirondack Mountains. 3. Human–animal
relationships—New York—Adirondack Mountains. I. Sharpe, Virginia A.
(Virginia Ashby), 1959– II. Norton, Bryan G. III. Donnelley, Strachan.
IV. Title.
   QL737.C22 W6475 2001
   333.95'9773'097475—dc21
                                        00-011113

Printed on recycled, acid-free paper

Printed in Canada
10 9 8 7 6 5 4 3 2 1

Contents

Acknowledgments

The chapters in this book grew out of a conference held in October 1998 at the American Museum of Natural History in New York City. We would like to thank Francesca Grifo, former director of the museum's Center for Biodiversity and Conservation, and her staff members Margaret Law and Meg Domroese for their assistance in planning that event and making available the conference space. We would also like to acknowledge Pam Davee of Georgia Institute of Technology, who worked as an intern on the project, and Hastings Center staff members Janet Bower, Marguerite Strobel, Ellen McAvoy, and Sandra Morales for their excellent technical support of the conference and the preparation of this volume. Thanks also go to the conference attendees and speakers and to all those who wrote chapters for this volume.

Both the conference and this volume were made possible by a generous grant from the National Science Foundation (grant number 97-12377). We thank NSF program official Rachelle D. Hollander for her support of the project and her ongoing support of interdisciplinary work on science and values. Finally, we would like to thank Dr. Leigh Turner for his role in drafting the successful proposal to NSF.

Introduction

Virginia A. Sharpe, Bryan G. Norton, and Strachan Donnelley

There is nothing simple about a story of wolves and human communities. It is a complex story with a rich history, an often controversial present, and a future that is not yet written. It is a story of human psychology and the ways in which our self-understanding, fears, hopes, interests, and sense of place shape our understanding of the natural world and our relationship to nonhuman species and the land. The story of wolves and human communities is also a story about our relationship to other human beings. The world we have today, at every level from the local to the global, is a world shaped by the choices of our forbears. For the short span of time that each of us will exist on the earth, our identities, decisions, and daily lives are shaped by our relationships with our families, our communities, our institutions, and those with whom we are at common or cross-purposes. The demands of our daily lives and our clear responsibilities at home and at work make it very easy for us to concentrate on our immediate present. But our relationships inevitably extend beyond the present. Just as our forbears have shaped our world, so will our choices affect the lives of our posterity.

The Endangered Species Act (ESA) is both an effect and a cause of this story's complexity. The act's passage in 1973 reflected a growing awareness in the United States and other nations of the dire consequences of the economic development and expansion of the nineteenth and twentieth centuries. Specifically, the act was designed to address the problem of species extinction and to reinforce the value of conservation in the nation's regulatory framework. Echoing the U.S. Constitution's intent to "secure the Blessings of Liberty to our selves and to our Posterity," the ESA identifies as one of its purposes the "better safeguarding, for the benefit of all citizens, the Nation's heritage in fish and wildlife" (§2.3). Among its requirements, the act stipulated that the federal government should "use all methods and procedures . . . necessary to bring any endangered or threatened species to the point at which the measures provided in this Act are no longer necessary" (§3.2). With this provision, the restoration of species became, for the first time, a requirement of U.S. public policy. The gray wolf was the first animal to be listed as endangered under the ESA, and it is under the auspices of the ESA that restorations of the gray wolf in Yellowstone National Park and central Idaho, the red wolf in North Carolina, and the Mexican gray wolf in Arizona and New Mexico have taken place.

In 1992, the U.S. Fish and Wildlife Service, the agency charged with managing restoration on federal lands, released a recovery plan for the eastern timber wolf identifying the Adirondack Park in northern New York State as one of several potential areas in the Northeast for restoring this subspecies. Because the Adirondack Park contains no federal land and is therefore under limited federal jurisdiction, the U.S. Fish and Wildlife Service would, at most, provide guidelines for the state Department of Environmental Conservation (DEC) to consider wolf recovery in the area. Because of higher program priorities and staffing and fiscal constraints, the DEC has taken no action on the feasibility of wolf restoration in the Adirondack region. Nevertheless, and probably because of the DEC's decision to table the matter, a proposal to restore wolves to Adirondack State Park has been introduced by Defenders of Wildlife, a national conservation organization.

Like wolf restoration activities in the West, this proposal has generated intense public debate on issues of property rights, land use, obligations to present and future generations, animal rights, wildlife management, biodiversity and the health of ecosystems, and natural recovery (where wolves return to an area on their own) versus reintroduction. In this volume, we have chosen the Adirondack restoration proposal as the basis for an exploration of these issues. Although the authors of particular chapters advance arguments for and against the Adirondack proposal, the book itself neither offers nor advocates a particular position on wolf reintroduction to this area. Rather, the premise of the book is that the question of what *should* happen is a question of political, social, cultural, ethical, and ecological values and the process by which these values are debated. By bringing together diverse viewpoints, we hope this book will contribute significantly to these debates.

Because Adirondack Park includes no federal land, the federal jurisdiction over any wolf reintroduction is limited to the federal interest in the transport and handling of the wolf as an endangered species. Otherwise, the decision will be a function of deliberation at the state and local levels within New York.

In our highly bureaucratic and technical society, there is a temptation to believe that questions such as the appropriateness or feasibility of wolf reintroduction can be answered solely by better or more objective scientific evidence. Essential though this evidence is, the values that different stakeholders bring to public policy debate are equally essential to the decision-making process. In other words, stakeholders bring their own expertise to policy decisions, expertise that may take the form of values about the kind of community they want to create, the relationship they want to have to nature and to human institutions, and the long- and short-term benefits and burdens they see in various policy options. In a democratic society, a legitimate public policy process will incorporate this array of considerations. Indeed, this is one of the foundations of the State of New York Environmental Quality Review Act (SEQRA) and the National Environmental Policy Act (NEPA) on which it is modeled. As in the federal regulations, the state mandates broad assessment of the environmental and human impacts of proposed actions deemed to have environmental significance. So, although one of the broad aims of this volume is to acknowledge and understand the importance of public participation in environmental management, the diverse perspectives gathered here also serve to forecast many of the issues that will emerge in any SEQRA-mandated environmental assessment of the specific Adirondack proposal.

A number of factors underscore both the challenge of and the need for a constructive deliberative process, First, unlike national parks such as Yellowstone that include no private land, the Adirondack Park is a 6-million-acre mix of state and private land that includes 105 townships and 130,000 permanent residents. The creation of the park in 1885 and the restrictions that the Adirondack Park Agency imposes on private land use are the source of considerable and long-standing resentment on the part of many Adirondack residents. Second, the historical tensions captured in charges of an upstate versus downstate perspective inevitably resurface when a proposal directly affecting a portion of the state's population is to be decided by citizens from the entire state. Third, a controversial issue such as wolf reintroduction has the potential to divide communities and individuals in the park who depend on one another to support the region's fragile economy.

These high stakes are raised even higher if we consider threats to ecosystem health and concerns about the kind of environment we will leave to our children. If, as the U.S. Constitution makes plain, we are to "secure the Blessings of Liberty for ourselves and our Posterity," how will we interpret and institutionalize the role of biodiversity in realizing this goal? Is restoring wolf populations sufficiently important in the larger picture of biological conservation to warrant special government efforts?

What biology tells us, Aldo Leopold reported in his metaphorical essay "Thinking Like a Mountain" (Leopold 1949), is that the value of wolves cannot be observed solely within the present and immediate; their value emerges at the scale of the mountain (Leopold's metaphor for the ecosystem) and in the longer reaches of ecological and evolutionary time: "Only the mountain has lived long enough to listen objectively to the howl of a wolf," Leopold concluded. By controlling populations of grazing animals, wolves protect the vegetative cover that holds the mountain together. Wolves are especially important because, as top predators, they set the constraints against which other species compete, closing the circle of evolution, retaining some places in nature where every species responds directly or indirectly to the forces of a complex ecosystem, including a top predator. As the carnivorous capstones of natural ecosystems, wolves represent the wholeness and the dynamism—the very creativity—of wild nature itself. It is against the constraints set by wolves that deer and other ungulate populations carve a niche in the complex process of ecosystem formation and development. The findings of biology thus tell us that the value of wolves should not be reduced to the individual or economic scale of time. Wolves are a part of our natural history at the scale of the ecosystem formation, a function that shapes the very physical and natural context in which we live.

If some subspecies of wolf is indeed the historical top predator in the Adirondack ecosystem and if species diversity is fundamental to maintaining freedom of choice for our posterity, do we not have an obligation to attempt to restore the integrity of that ecosystem by restoring wolves? Likewise, if we see it as our responsibility to restore the top predator to an ecosystem, does our responsibility end here or, rather, extend to management so that wolves and humans can coexist indefinitely, thus reestablishing some sort of harmony between human and natural communities?

All these considerations and the prospect of a state and local deliberative process regarding wolf reintroduction encouraged us to plan and convene a conference on Wolves and Human Communities in October 1998. At that conference, which was sponsored by the National Science Foundation and held at the American Museum of Natural History in New York, approximately 25 biologists, policymakers, politicians, philosophers, sociologists, lawyers, and Adirondack residents made the presentations that are the bases for the chapters in this volume. We have divided the book into seven sections that address the issue of wolf restoration from roughly the specific to the general, the present to the future, the local to the global. The themes we have identified in this introduction are woven together throughout the chapters of the book. Thus, although the authors are writing from diverse perspectives, they often draw on common resources to make their case. To avoid repetition, these common resources are gathered at the end of the book. Where overlap occurs between chapters, we take this to be a signal of a common theme that transcends the often divisive and controversial question of wolf reintroduction.

Part I opens the book with three chapters that highlight the complexities of any public process regarding wolf reintroduction in the United States. The chapters in Part II present the views of stakeholders in the Adirondack decision as well as a chapter on historical trends in public perception of restoration generally. Part III focuses on the legal and policy context for species preservation and explains how restoration challenges the system of property law that has evolved over centuries in the Anglo-American tradition. Part IV examines biological and political lessons learned from Yellowstone, Isle Royale, and the Great Lakes states. The three chapters in Part V invite us to reflect on the meaning of wildness, in both ourselves and the wolf, and the threat and promise wildness offers to our sense of human responsibility. Part VI examines the implications of biodiversity for long-term human obligations. Finally, taking the point of view of evolutionary time and ecological scale, Part VII challenges us to develop a new consciousness about our position in the natural world.

Unlike natural restoration, in which a species finds its own way back to a landscape it once occupied, species reintroduction is brought about by conscious decisions and actions on the part of humans. What exactly do we need to be conscious of as we consider the question of wolf reintroduction? The different perspectives offered in this volume are an effort to give this question its broadest possible answer.

Part I

Place and Process: Political and Policy Perspectives on Wolf Restoration

We open this volume with three chapters that consider the necessary elements of a constructive public process on the question of wolf reintroduction. Daniel Kemmis, former mayor of Missoula, Montana, four-term Montana legislator, and director of the Center for the Rocky Mountain West, identifies respect as one of the most fundamental ingredients.

In his evocative Chapter One, Kemmis offers a meditation on the sovereignty of charismatic megafauna, the meaning of place, and the need for respect for wild nature and the dignity of human communities. Kemmis argues that respect for the power of the wolf as a charismatic species entails a parallel respect for the sovereignty of citizens in local and regional governance. Although he does not offer specific proposals on the shape new collaborative forms of governance should take, especially as they bear on complex ecological questions, he makes it clear that a sustainable human future cannot be achieved if individuals and communities are disempowered in the political process. The two chapters that follow make some concrete observations on this call for respectful collaboration in wolf reintroduction.

In Chapter Two, L. David Mech, senior scientist with the U.S. Department of the Interior Biological Resources Division, internationally known wolf expert, and founder of the International Wolf Center in Ely, Minnesota, describes the pitfalls and promise of public participation in wolf reintroduction, with specific attention to wolf advocates.

Citizens have played a positive role in wolf restoration decisions by voicing their preferences and creating compensation programs responsive to wolf depredation. The negative influence of the public is seen in misrepresentation and misinformation by advocacy groups that promote their agendas without regard for documented evidence.

Central to a candid policy debate, Mech argues, is the need to be clear at the outset that any biologically successful wolf reintroduction includes constant management of growing populations; some wolves must be killed as part of ongoing effective management. Because wolf recovery often provides an opportunity for advocating disparate agendas, Mech observes that without explicit clarification of and deliberation about the values at stake, the use of reliable scientific data, and honesty about the controversial issue of wolf management, partisan controversies will discourage politicians from taking up the question of wolf reintroduction.

In Chapter Three, Robert Inslerman, wildlife manager in the Adirondack region for the New York State Department of Environmental Conservation (DEC), describes the topography, history, and demography of the park and the DEC's role in the proposed wolf recovery. Emphasizing the political need for feasibility, Inslerman explains that DEC will entertain a restoration proposal only if it is demonstrated that an informed public supports or at least does not oppose the release of wolves. Public support, or lack thereof, will be ascertained by DEC through a public process including a draft environmental impact assessment, as required by the State Environmental Quality Review Act.

Inslerman lays out the array of issues that must be addressed as part of the environmental impact assessment. These include the historical status of wolf subspecies in the Adirondacks, including a genetic analysis of the eastern coyote *(Canis latrans)* that currently inhabits the park; the biological feasibility of reintroduction in terms of sufficient land and prey and the potential for hybridization between coyotes and wolves; the ecological impacts on beaver, deer, and the landscape; the impacts on the human communities in the Adirondacks, including issues of land use, hunting and trapping, safety, and local economics; the relative weight and authority of interests within and outside the park; and the immediate and long-term costs to DEC associated with the release and management of wolf populations.

These chapters sound this book's central theme of democratic participation as a fundamental basis of wolf restoration.

Chapter One

Wolves as Bioregional Sovereigns

Daniel Kemmis

My observations are those of a practical politician who is increasingly challenged by but optimistic about our ability to deal with complex wildlife restoration issues. To deal with these issues in a long-term, sustainable, effective way, we must challenge our thinking about what it means to be human in such situations and what it means to make these kinds of decisions together as human beings. It is against this backdrop that I would like to talk about wolves.

We might examine first the special nature of restoration or preservation issues as they relate to charismatic megafauna, the high-profile species of which wolves are a classic example. (Other examples include grizzly bears, eagles, salmon, tigers, and mountain lions.) I want to examine these charismatic megafauna and the special challenges their restoration and preservation pose for us. How can wolves teach us something about our own humanity and about how we as humans make decisions and create policies about other species?

Consider the phrase *charismatic megafauna*. What gives certain species this charisma? I maintain that wolves or grizzly bears, eagles or elk, move us because of our instinctive response to the sovereignty we sense in them. Some may view this as anthropocentrism, but I ask that you hold such objections in abeyance for the time being. Certainly we can agree that something about these charismatic creatures claims our attention differently than, say, a snail darter does. Let's ponder further what that "something" might be.

Our response to the sovereignty of these animals can serve as a starting point for us as we struggle with what it means to be responsible not only to them but to all creatures. When we feel this mysterious instinctive response, we might look for the stirrings of a new, deeper, and more reliable understanding of what such responsibility might be about and how it might best be exercised.

This raises two questions. First, what are we actually responding to in these creatures? Second, what form should our response take? To address the first question, what we're responding to seems to be a matter of form. Something

9

about the form of these creatures moves us profoundly. Consider the form of the wolf. What is it about the wolf that is so commanding? When we contemplate the physical being of a great wolf, we can't help but think of its power, its grace—indeed, how highly evolved in form this particular creature is. In recognizing that something about the form of the wolf moves us, we can't avoid hearing echoes of our own human nature. We are so moved by the wolf because we ourselves are such a highly evolved species. We also are a powerful species; we also are capable of being amazingly graceful. A large part of why wolves are important to us, then, is that they reflect our own being back to us. As soon as we start examining what evokes our response to the wolf, we start learning something about ourselves.

Furthermore, when we look at the wolf, we recognize the tremendous work evolution has been carrying out for so many millions of years. However highly evolved our species is, we recognize that the creation of such forms is work that we are incapable of doing. As a species proud of our own creativity, and rightfully so, we naturally respond with respect or awe to an even greater creativity and greater potency than we possess.

There is something else that we respond to in the wolf because it is so important to the existence of our own species. Consider how the wolf has evolved into its specific form, not just the physical form of the individual wolf but also the form of the wolf pack, the social form of wolves. It is impossible to consider these evolved forms without becoming aware of a very powerful relationship between the wolf and its place. The form and shape of the wolf— what it is as a being—is fundamentally a response to its place. Its howl, for example, is a response to the place, to the distances across which a pack must range. So is the development of the powerful muscles and amazing endurance of the wolf; these too are ways of being in a certain place. And here we begin our return to the idea of sovereignty.

Why might we be inclined to call the wolf a sovereign creature? If we take apart the word *sovereign,* we find another word that turns out to be of tremendous significance to many human activities, including species recovery efforts. *Sovereign* contains the word *reign,* which in turn reflects *region;* the Latin root *regere* is the same in both. The wolf is a creature of region: It is inseparable from its region, unknowable apart from its realm. We simply cannot understand how the wolf came to be, or how it can continue to be, except in relation to the region in which it reigns. So here are the beginnings of a reflection on some fundamental aspects of our own human nature and of how it affects the way we approach issues such as wolf recovery. In dealing with issues of recovery or restoration of any of the charismatic megafauna, we are inevitably brought to questions of region and regionalism. We find ourselves faced with the question of how we, as human beings, exercise our own sovereignty, and specifically of the relationship of human sovereignty to meaningful regions.

Dealing with habitat preservation issues in a sustainable way entails sorting out how we, as humans, divide up our rule, our sovereignty, over the landscape. Inevitably when we deal with these tough issues, we are drawn into a dialogue about who is in charge of the landscape. Specifically, we are drawn more and more into conflicts between national sovereignty and regional, localized sovereignty. An example from my own bioregion in the northern Rockies has to do with grizzly bears and their possible reintroduction into the Selway-Bitterroot region, which bridges the state line between Montana and Idaho.

According to the U.S. Fish and Wildlife Service, Selway-Bitterroot is one of the few areas in the Lower 48 where grizzly reintroduction would be feasible. As the Fish and Wildlife Service explored potential grizzly recovery in that area (which includes a number of wilderness areas and national forests), a variety of recovery plans emerged. The preferred alternative was the result of a collaboration between local environmentalists (including the local Defenders of Wildlife and National Wildlife Federation branches) and industry representatives (various timber producers and sawmill unions). Together they devised a proposal for an initially small but steadily growing experimental population of grizzlies, similar to the wolves recently reintroduced under this designation in Yellowstone National Park. The bears would be managed by a citizens' committee appointed by the governors of Montana and Idaho.

This proposal was met with skepticism, especially from national environmental groups, who argued that the land in question, being national forest land, falls fully and clearly under the jurisdiction of the national government. What right do the people in the northern Rockies have to come up with their own plan for managing these national forests, let alone to put locals in charge of a federally designated threatened species?

This is an issue of sovereignty, a very important issue that will not disappear but will become more insistent and challenging as we move forward into this complex arena of ecosystem management and biodiversity preservation. These sovereignty issues are bothersome, but in some way they may be good for us; they force us to ask some long-suppressed questions about ourselves. How is it most likely that we, burdened or blessed as charismatic megafauna, in our own right a sovereign species, will rise to the challenges that face us?

At the very least, we must address one another in the midst of these controversies with as much respect for our own species as for the animals we are seeking to protect. This means, in the case I cited, that westerners of different political persuasions must be able to look each other in the eye, confronting together the challenges of inhabiting that Rocky Mountain landscape and finding workable solutions together. As a practicing politician, one who has always been a democrat and who happens also to be a Democrat, I am convinced that we must begin to develop forms of government that allow us to work out and implement solutions with our neighbors without top-down, paternalistic rules and regulations.

But that takes some new approaches to sovereignty. When national environmental groups objected to the grizzly plan on the basis that "these are national forests," the regional forester in Missoula-based Region I responded that although the lands in question certainly are national forests, there are and always will be endless opportunities for locals to undermine and sabotage any centrally devised and imposed recovery. If the locals don't own the plan, it can't work. Although sovereignty in a strictly legal sense may reside in Washington, unless the people of the region in question are given a way to exercise meaningful sovereignty, the wolves or the grizzlies will exercise no sovereignty at all.

I conclude that we can succeed at protecting sovereign species only by developing some new forms of local and regional sovereignty. Eventually, though, what we must confront in species preservation is our global citizenship and our capacity for global sovereignty. And this challenges in a new way the forms of sovereignty that we have already evolved. However attached we may be to paternalistic approaches at the national level, paternalism fails on the global scale. No one outside the earth (at least no one of our own species) is capable of imposing a paternalistic system that will guarantee that we take good care of the earth. Therefore, if the earth is going to take care of itself, if we as citizens of the earth are to care for it and rise to the global challenges that face us, we must do so as a democratic people, as a truly sovereign people. But the only way we can learn to fulfill this duty of global citizenship is to exercise full democratic responsibility in the places we inhabit. Unless we learn how to give the people of the northern Rockies or the Adirondacks a full range of responsibility, a full sovereignty over their own regions that matches the sovereignty of the wolf, we will never be able to fulfill our responsibility to the wolves themselves.

Chapter Two

Wolf Restoration to the Adirondacks: The Advantages and Disadvantages of Public Participation in the Decision

L. David Mech

The first time I ever saw a wolf in New York State's Adirondack Mountains was in 1956. It was a brush wolf, or coyote *(Canis latrans),* not a real wolf, but to an eager young wildlife student this distinction meant little. The presence of this large deer-killing canid let my fresh imagination view the Adirondacks as a real northern wilderness.

Since then I have spent 40 years studying the real wolf: the gray wolf *(Canis lupus).* Although inhabiting nearby Quebec and Ontario, the gray wolf still has not made its way back to the Adirondacks as it has to Wisconsin, Michigan, and Montana. Those three states had the critical advantages of nearby reservoir populations of wolves and wilderness corridors through which dispersers from the reservoirs could immigrate.

The Adirondacks, on the other hand, are geographically more similar to the greater Yellowstone area in that they are separated from any wolf reservoir by long distances and intensively human-developed areas aversive to wolves from the reservoir populations. If wolves are to return to the Adirondacks, they almost certainly must be reintroduced, as they were to Yellowstone National Park.

Wolf reintroduction, as distinct from natural recovery, is an especially contentious issue, for it entails dramatic, deliberate action that must be open to public scrutiny, thorough discussion and review, and highly polarized debate. This is as it should be because once a wolf population is reintroduced to an area, it must be managed forever. There is no turning back. The wolf was once eradicated not just from the Adirondacks but from almost all of the 48 contiguous states. That feat was accomplished by a primarily pioneering society that applied itself endlessly to the task, armed with poison. We can never return to those days, so once the wolf is reintroduced successfully, it will almost certainly be here to stay.

The Wolf as an Endangered Species

For those who have not followed wolf reintroductions elsewhere or who think of the wolf as just another endangered species, it is useful to emphasize that unlike most other endangered species, the wolf was deliberately exterminated for a reason.

Perhaps comparing the wolf with Kirtland's warbler *(Dendroica kirtlandii)* would be instructive. Kirtland's warbler, a tiny bird of little notice to most people, became endangered when its habitat was destroyed incidentally to other human activities. Kirtland's warbler occupied a very narrow niche in that it needed a certain age of jack pine forest in which to nest. When placed on the endangered species list, Kirtland's warbler numbered only a few individuals, and the species was not inimical to human activities.

The wolf, on the other hand, was exterminated deliberately because of its depredations on livestock. These depredations fostered a general fear and loathing toward the wolf that resulted in the animal being controlled, persecuted, and wiped out even in areas where it was not doing damage to livestock, such as Yellowstone National Park and other parts of the western wilderness (Young and Goldman 1944). Furthermore, the wolf was familiar to the public, although most people knew the animal primarily through myths, legends, and fairy tales. When the wolf was placed on the endangered species list, that endowed the animal with a new image.

The Endangered Species Act (Endangered Species Act 1973) came at a time when the public was first becoming widely aware of environmental concerns, a period so profoundly important that I have often called it the Environmental Revolution (Mech 1995a). To the public newly imbued with environmentalism, the wolf became a symbol of endangered species.

However, not all endangered species were equal in their degree of endangerment. Whereas species such as Kirtland's warbler were endangered worldwide and therefore were truly on their last legs, the wolf was endangered only locally. Some 100,000 to 200,000 still lived in Canada, Europe, and Asia. This distinction was lost on many people and ignored by others, who saw treatment of the wolf as symbolic of the human assault on the earth.

Wolf Subspecies

The Endangered Species Act defined endangered species as either species or subspecies. For the wolf, this provision confused the public greatly, for originally there were 24 recognized wolf subspecies in North America (Young and Goldman 1944; Mech 1970). This contrasts greatly with the number of subspecies recognized in Europe and Asia, which was only 8. The difference between wolf taxonomy in the Old and the New World boiled down to the tendency on the part of New World taxonomists to split rather than lump wolf subspecies. Thus some of North America's wolf subspecies were identified based on a single skin or skull.

The whole concept of subspecies is a contentious one deriving from the human attempt to categorize natural differences between organisms. On the subspecies scale, the distinctions have been very subjective and questionable in value (Wilson and Brown 1953; Mayr 1954).

In the case of the wolf, this confusion was addressed in Nowak's (1995) revision of wolf taxonomy in North America. Nowak lumped the 24 subspecies into only 5. Because of the wolf's great mobility (Van Camp and Gluckie 1979; Fritts 1983; Ballard, Farnell, and Stephenson 1983; Gese and Mech 1991; Boyd et al. 1995), it was clear that the North American wolf population was mixing greatly, thus tending to refute the concept that the population comprised large numbers of physically different groups of wolves. Nowak's 5 subspecies made much more sense than the original 24.

The U.S. Fish and Wildlife Service responded to the Endangered Species Act by setting up a team for each species on the list to plan their recovery. In the case of the wolf, teams were established at first for each of the 24 wolf subspecies that may still have had members remaining in the 48 contiguous states, including the eastern timber wolf *(Canis lupus lycaon)*, the Mexican wolf *(C. l. baileyi)*, and the northern Rocky Mountain wolf *(C. l. irremotus)*. In addition, the red wolf *(Canis rufus)*, a taxonomic enigma to this day (Nowak et al. 1995), was granted its own recovery team. The teams were made up of experts on the wolf and administrators from various state and federal agencies who would be involved in the recovery of the specific animal.

By 1978 the U.S. Fish and Wildlife Service had changed the way it listed wolves as endangered. All remaining wolves in the 48 contiguous states were listed, rather than the individual subspecies. Thus, regardless of subspecies, wolves anywhere in the 48 states were on the list. However, the recovery teams for each subspecies retained their names, and the plans they developed retained the names of the subspecies.

Nowak's (1995) taxonomic revision necessarily redrew the lines separating the subspecies, and his data showed that the eastern race of wolf, the eastern timber wolf *(C. l. lycaon)*, had inhabited the northeastern part of the United States east of Michigan. Under the pre-1995 classification, however, the eastern timber wolf range had extended almost to the western border of Minnesota, thus including Michigan, Wisconsin, and most of Minnesota in its range.

However, the only remaining wolves in the eastern United States when the animal was placed on the endangered species list were those in Minnesota and on Lake Superior's Isle Royale (a part of Michigan). Thus the Eastern Timber Wolf Recovery Team, which had examined the status of this race of wolf in all its former range, had decided that the best region in which to promote wolf recovery was Minnesota, Wisconsin, and Michigan.

There were several reasons for this decision. First, wolves had inhabited Michigan and Wisconsin until the 1960s (Hendrickson et al. 1975; Thiel 1993). Second, both northern Michigan and Wisconsin included large amounts of wilderness or semiwilderness in public holding, and Minnesota still harbored several hundred wolves (Cahalane 1964; Mech 1970). Third, there was reason to believe that wolves from Minnesota could disperse into Wisconsin and Michigan (Mech and Frenzel 1971). Fourth, there was confusion about whether the wolf that occurred originally in the southern Appalachians was the eastern timber wolf or the red wolf. Finally, state agencies in Maine and New York had responded negatively to queries by the Eastern Timber Wolf Recovery Team as to whether they would consider restoring wolves to their states (U.S. Fish and Wildlife Service 1978).

The Eastern Timber Wolf

Nowak's (1995) revision redesignated the wolf in Minnesota, Wisconsin, and Michigan as the Great Plains wolf *(C. l. nubilis)* rather than the eastern timber wolf *(C. l. lycaon)*. This meant that the eastern timber wolf recovery plan (U.S. Fish and Wildlife Service 1978, 1992) had set criteria for recovery of the Great Plains wolf and that there was no plan for recovery of the eastern timber wolf. This was not a problem from a legal standpoint because the wolf was no longer listed by subspecies. Nevertheless, it allowed the public to claim the need to restore the eastern timber wolf, which still survived in a reservoir in southeastern Ontario and southern Quebec.

Whether the federal government is obligated to attempt to recover the eastern timber wolf is not clear. Just how far the government must go in recovering an endangered species has not been tested. Each wolf recovery team has made its own recommendations as to when the population in its purview can be considered recovered. However, the question of how many wolf populations must be recovered to accomplish the goal of wolf recovery under the Endangered Species Act has not yet been resolved.

Nevertheless, because the Eastern Timber Wolf Recovery Team considered whether wolf recovery was necessary in the Adirondack Mountains and decided that recovery in Wisconsin and Michigan would suffice, the federal government seems justified in agreeing that wolf recovery in the Adirondack Mountains is not integral to wolf recovery nationwide.

This decision does not mean that the federal government would oppose wolf recovery in the Adirondacks. In fact, the U.S. Fish and Wildlife Service is developing a recovery plan for the wolf in the northeastern United States. This plan would provide guidelines for any state that wanted to promote wolf recovery in the area (R. Refsnider, personal communication,1998). Whether wolf recovery in the Adirondacks will become a high priority for the Fish and Wildlife Service or whether it would even be funded remains to be seen. Furthermore, New York State could elect to restore the wolf on its own, just as any other state could do.

Advantages of Public Participation in Decision Making

This is where public involvement can play a very important role. If enough citizens in New York want their government to restore wolves to the Adirondacks, such a measure will have to be given a great deal of consideration, at least by the state. Or if enough U.S. citizens favor wolf restoration to the Adirondacks and inform their elected representatives, then as long as the wolf remains on the federal endangered species list, the federal government will have to listen.

On the other hand, if enough public pressure is brought against the issue, then that will have considerable influence. Because of the wolf's controversial nature and its tendency to prey on livestock and pets, there will be public opposition. What the net balance of public sentiment will be and how that feeling will be viewed by representatives and agencies remain to be seen. However, it seems likely that without strong positive public pressure, wolves probably will not be restored to the Adirondacks.

This illustrates one of the main advantages of public pressure. The government has many tasks to accomplish with endangered species and far too few personnel and too little funding to deal with all of them as they deserve. In prioritizing those tasks, the government must consider public interest and pressure. Bringing this pressure is the most significant role the public can play in wolf recovery in the Adirondacks. In fact, it was just such public involvement that helped precipitate wolf reintroduction into Yellowstone National Park in 1995 and 1996.

Public involvement also has assisted wolf recovery through the payment of compensation for wolf depredations on livestock in the Rocky Mountains (Fischer, Snape, and Hudson 1994). This program was initiated by the Defenders of Wildlife and parallels a program by the State Department of Agriculture in Minnesota (see Schlickeisen, Chapter Seven, this volume). What is so innovative about the Defenders program is that it allows wolf advocates to help subsidize their favorite animal rather than making the general public pay, as government programs would. There is little doubt that the Defenders compensation program greatly facilitated the reintroduction of wolves to Yellowstone and central Idaho, and it can be argued that without the Defenders program, reintroduction might not have taken place. Similarly, the Defenders compensation program greatly expedited reintroduction of the Mexican wolf into Arizona.

Groups that deserve special mention in promoting wolf reintroduction into Yellowstone are Renee Askins's Wolf Fund, which was established solely to promote wolf reintroduction into Yellowstone, and Bobbie Holaday's Preserve Arizona's Wolves (PAWS), which lobbied primarily for the Mexican wolf reintroduction. Both organizations disbanded once the goal of reintroduction was met. Several other organizations also lobbied strongly for these reintroductions and for wolf recovery in general.

Disadvantages of Public Participation in Decision Making

Organizations promoting wolf recovery or protection come and go. At any given time there have been as many as 51 such organizations (International Wolf Center 1982). Most were founded by laypeople, although a few employ biologists as staff or consultants. Furthermore, most of them actively advocate for the wolf through lobbying or urging members to lobby legislators and government administrators, disseminating information (some of which may not be supported by available research), and writing letters to newspapers. Only one, the International Wolf Center in Ely, Minnesota, professes to advocate for the wolf without these tactics, relying entirely on disseminating objective, accurate information about wolves.

Thus the quality and accuracy of the information disseminated by these organizations varies widely. Some admit deliberately disseminating false information (Anonymous 1992), whereas others do so out of naivete, carelessness, or failure to check information obtained from other groups.

The misinformation promulgated by wolf advocacy groups ranges from minor technical errors to major deception and fraud (Blanco 1998). Technical biological misinformation, though bothersome to professionals working with wolves, is not as serious as deception about such issues as the status and trends in wolf populations. This latter type of misinformation tends to motivate well-meaning wolf advocates to press their causes through letter-writing campaigns, public meetings, lobbying, and lawsuits. For example, animal welfare and wolf advocacy groups have been advertising for funds in major national newspapers for years, claiming that wolves were threatened in Denali National Park and other parts of Alaska (*USA Today* 1995, 1998), despite documentation to the contrary (Stephenson et al. 1995; Mech et al. 1998).

These misrepresentations have even made it into conference proceedings. In the non–peer-reviewed proceedings of a nonprofit citizen organization, "Defenders of Wildlife's Restoring the Wolf Conference," undocumented claims were made that the wolf has been eliminated from "95% of its former range" and "95% of its historic range in North America" (Valentino 1998, 47–48). The actual figures are closer to 30% of its global range and 40% of its North American range (Mech 1970; Ginsberg and Macdonald 1990).

In the same proceedings, it was alleged that "wolf populations throughout the state [of Alaska] are being decimated, including those in our national parks and preserves" (Joslin 1998, 90). This conclusion contrasts dramatically with the peer-reviewed, documented conclusions that Alaska's 5,900 to 7,200 wolves have long been well managed (Stephenson et al. 1995) and that in Denali National Park and Preserve, humans kill less than 4% of the wolf population annually (Mech et al. 1998).

A current case involves the issue of removing the wolf from the federal endangered species list in Minnesota, Wisconsin, and Michigan. Wolf recovery there has been well planned, methodical, and well documented. Wolf recovery in Minnesota, Wisconsin, and Michigan was the subject of one of the federal government's first recovery plans. The planning process involved many months of meetings, considerable public input, and several revisions based on new information and on changes in the wolf population (U.S. Fish and Wildlife Service 1978, 1992). Detailed criteria were set for the size of wolf populations to be required for formal recovery and the delisting of that wolf population from the endangered species list.

The wolf population in the area at the time included about 650 wolves in Minnesota and none in Wisconsin or Michigan except for those on Isle Royale (Mech 1970). The recovery plan recommended that there be at least 1,250 wolves in Minnesota and 100 in mainland Wisconsin and Michigan combined for 5 consecutive years, and there was little or no dissent by wolf advocates when these figures were proposed (U.S. Fish and Wildlife Service 1992). As the populations recovered and the population goals were met, however, wolf advocacy groups began opposing the proposed delisting.

Despite the well-documented, thorough, peer-reviewed population estimation procedures (Fuller et al. 1992; Berg and Benson 1999), the wolf advocacy groups claimed, without documentation, that the resulting estimate of 2,450 wolves in Minnesota was too high (Berg, Hatfield, and Brave Heart 1998). Defenders of Wildlife (1998) and other groups maintained that delisting the wolf in Minnesota would be premature.

This claim of premature delisting was made despite the following facts: There are three to four times as many wolves in Minnesota today (Berg and Benson 1998) than probably any other time in roughly the last 100 years (Herrick 1892; Surber 1932), the wolf population has been increasing at an annual average of 3% to 5% since 1979 (Fuller et al. 1992; Berg and Benson 1998), they occupy 46% more range than in 1989 (Berg and Benson 1998), a majority of Minnesota citizens favor wolf recovery (Kellert 1986; Minnesota Poll

1998), and no threat to the well-studied Minnesota wolf population has been identified.

In Spain, an even more extreme situation has developed. According to Juan Carlos Blanco, who has used radio tracking and other techniques to study the wolf in Spain (Blanco 1998; Blanco, Cuesta, and Reig 1990), "radical environmentalists" published in a newspaper results of a fraudulent wolf census that claimed serious declines in wolves. "The information, which has been picked up by all the newspapers, is totally untrue, made up" (Blanco 1998, 2).

Such tactics on the part of extreme wolf advocates are seized upon by wolf opponents as indicative of duplicity and extremism by environmentalists and as reasons to mistrust wolf advocates and wolf recovery or reintroduction efforts. Meanwhile, the views of the majority of the public, who would accept wolves in moderate numbers (Minnesota Poll 1998), receive little attention.

Thus public attitudes polarize, fostering discord and dissension. It is only human nature for politicians and busy bureaucrats to try to avoid situations that promise to be time-consuming and fraught with public dissatisfaction (Mech 1995a). For example, a proposal to reintroduce red wolves to the Land-Between-the-Lakes area in Kentucky and Tennessee was withdrawn because of adverse public opinion (Parker and Phillips 1991).

Blanco (1998, 3) in Spain stated the problem of extreme wolf advocacy as follows: "The worst effect of this campaign is that it will act against the wolf like a time bomb. We will see the most damaging effect within the next few years when the nature conservation movement becomes discredited in the eyes of Society. What will those members of the public who are genuinely concerned about the wolf's future think when they learn that the fall in wolf numbers in northern Spain is nothing but a huge lie?"

Restoring the Wolf to the Adirondacks

As mentioned earlier, restoring a wolf population should be considered a permanent act that will forever entail intensive population control. Originally it was only through the use of poisoning, den digging, bounties, aerial hunting, steel trapping, and snaring that wolves were eradicated. Poisoning is now illegal, and public distaste for many of these other techniques is high and increasing. Therefore, once a wolf population is reestablished, eliminating it will be very difficult and expensive (Mech 1998).

Meanwhile, a restored wolf population will continue to proliferate and colonize new areas as long as there is adequate prey. Each pair of wolves produces an average of five to six pups per year (Mech 1970), and within 2 to 3 years most offspring disperse over distances of up to 886 kilometers (Fritts and Mech 1981; Fritts 1983; Messier 1985; Mech 1987; Fuller 1989; Gese and Mech 1991; Mech et al. 1998). Thus in New York State, even the average minimum dispersal distance of 77 kilometers (Gese and Mech 1991) would take wolves from the Adirondack Park to half the state, including within easy colonizing distance of Catskill Park.

Because deer are among the main prey of wolves and deer are common in New York and surrounding states (Mattfeld 1984), wolves can be expected to spread quickly and colonize most of the rest of the state unless controlled. Wolf populations in Michigan and Wisconsin increased at average annual rates of 38% to 40% (Michigan Department of Natural Resources 1997; *Wisconsin Draft Wolf Management Plan* 1998). To control a wolf population, 30% to 50% of the wolves must be killed by humans each year (Mech 1970; Peterson, Woolington, and Bailey 1984; Fuller 1989; Ballard et al. 1997).

It is true that many areas of New York would benefit from wolves' controlling deer numbers, for deer can be pests to gardeners, orchardists, and other agriculturalists and cause millions of dollars of damage to vehicles, not to mention human injury and death. However, most areas of New York State outside the Adirondacks support many farms with livestock and pets. Wolves prey on both, and although the proportion of livestock killed may be low, such depredations cause ill feelings in the rural community (Fritts 1982; Fritts et al. 1992; Fritts and Paul 1989). Furthermore, fear of wolf attacks on children is a concern (Minnesota Poll 1998), especially because instances of this behavior were documented recently in India (Jhala and Sharma 1997; Cook 1997).

The Need for Wolf Control

In almost every area of the world that wolves inhabit, except parks and preserves, they are controlled, and any population restored to New York State would also need constant control. However, many of the strongest public advocates of wolf restoration oppose wolf population control (Berg, Hatfield, and Brave Heart 1998).

This dilemma faces each land or resource management agency administrator who considers wolf restoration. The result is a series of meetings, public hearings, consultations, environmental impact statements, legislation, and other types of red tape that greatly prolong the process. Thus it took more than 20 years to restore wolves to Yellowstone National Park (Fritts et al. 1995).

Much has been learned from previous wolf reintroduction efforts, and if restoring wolves to the Adirondack Park were merely a biological issue, the task could be done in a few years. An integral part of the reintroduction plan would be a plan to control the wolf population and minimize wolf depredations on livestock outside the restoration zone, specifying who will pay the substantial costs (Mech 1998).

Because of the aforementioned problems resulting from public participation in the wolf restoration decision, the question of whether wolves will be restored to the Adirondacks depends primarily on the tenor of public input. When it becomes widely publicized that after wolf restoration, hundreds of wolves will have to be killed each year to limit the population (Mech 1995a, 1998), even some wolf advocates will oppose restoration.

Whether to restore wolves to the Adirondacks ultimately will be decided by government officials. Anyone favoring wolf restoration to the Adirondacks would be well advised to heed the past and understand that a rational, scientific, professional approach probably is the only one that has any chance of winning official approval. If responsible public advocacy for wolf restoration in the Adirondack Park prevails, it is possible that when I again venture to that area, I may look out over a frozen lake and this time see a real wolf.

Chapter Three

Wolf Restoration in the Adirondacks: The Perspective of the New York State Department of Environmental Conservation

Robert A. Inslerman

In 1992, the U.S. Fish and Wildlife Service (USFWS) released a revised recovery plan for the eastern timber wolf *(Canis lupus lycaon)* that identified the Adirondack Park in northern New York as one of several areas to be investigated for eastern timber wolf reestablishment (U.S. Fish and Wildlife Service 1992). Restoring wolves to New York, if it occurs at all, must be a carefully thought-out, deliberate process that considers biological, social, political, and economic factors. Key to the success of any wildlife restoration program is public acceptance, particularly on the local level. In New York there are many supporters of wolves and wolf restoration, but there are also many opponents, whose opinions are equally strong, and they are important in the debate over whether to restore wolves to the Adirondacks.

The Adirondack Park

The Adirondack Park was created by an act of the legislature in 1885. Today it consists of a mixture of public and private land totaling approximately 6 million acres, about the size of Vermont (Viscome 1992). There are no check-in points, no toll houses or ticket booths where one has to register, no checkpoints where one has to stop when leaving. The park boundary is not fenced, blazed, or posted. The boundary is designated on most maps by a blue line known as the Blue Line. Large identification signs on major highways indicate when one enters the park, but lands inside the park look the same as lands immediately surrounding it. It is estimated that there are 130,000 permanent and nearly 200,000 seasonal residents in the 12 counties and 105 townships in the park. Nearly all forested, the park ranges in elevation from approximately 100 feet above sea level on Lake Champlain to more than 5,000 feet on Mt. Marcy, New York's highest point. Seven major ecological zones make up the park and surrounding area. Of the 6 million acres within the park, approximately 2.6 million (roughly 43%) are in public ownership and constitute the Forest Preserve. Protected by Article XIV of the 1894 state constitution, "The lands of the state, now owned or hereafter acquired, constituting the Forest Preserve as now fixed by law, shall be forever kept as wild forest lands. They shall not be leased, sold or exchanged, or be taken by any corporation, public or private, nor shall the timber thereon be sold, removed or destroyed" (New York State Constitution 1987). It is the only area in the country protected by a special provision in a state constitution. The remaining private lands, approximately 3.4 million acres (57%), together with the state lands, are governed by some of the strictest land use regulations in the United States. The Adirondack Park is a land of diversity and controversy: diverse in its weather, land uses, topography, and habitats; diverse in its inhabitants, human and wild; and diverse in the beliefs and customs of both residents and nonresidents, who will fiercely defend their opinions on land use, acquisition of additional land, wilderness protection, property rights, and the future use and protection of the Adirondack Park.

A History of Support

The Department of Environmental Conservation (DEC) has always supported restoration of native species. As stewards of the state's wildlife, DEC has made perpetuating extant native species and restoring extirpated ones an important part of its program since its foundation more than 100 years ago. It was one of the prime motivations for DEC's creation, one that continues to this day. It is the cornerstone of the Bureau of Wildlife's mission to enable the people of New York to enjoy all the benefits of the wildlife of the state, now and in the future. Wild turkey *(Meleagris gallopavo)*, bald eagle *(Haliaeetus leucocephalus)*, and peregrine falcon *(Falco peregrinus)* are some of the native species whose return was greatly accelerated by DEC's efforts.

Other projects have met with less success. State and private groups tried to restore elk *(Cervus elaphus)* and moose *(Alces alces)* several times between the late 1800s and the early 1900s, but all attempts failed. Attempts to establish the gray partridge *(Perdix perdix)* in the late 1920s and early 1930s met with questionable success. A university-sponsored release of lynx *(Lynx lynx)* in the Adirondacks in the late 1980s had uncertain results. Because of strong local opposition, DEC decided against restoring moose in the Adirondacks in the early 1990s.

A number of restoration initiatives are under way or being considered, each with varying degrees of involvement by DEC. In a partnership with the River Otter Project, Inc., river otter *(Lutra canadensis)* restoration in western New York is under way and appears to be succeeding. DEC is participating actively in this restoration effort. The Rocky Mountain Elk Foundation, through contracts with two New York universities, funded studies to determine the social and biological feasibility of restoring elk to suitable areas throughout New York. DEC assisted in this effort by listing the kinds of issues that had to be addressed before it could approve a proposal to restore and issue the necessary permits. A multistate partnership effort to restore trumpeter swans *(Olor columbianus)* to historical habitats is in the earliest stages of consideration, and DEC is involved in this process. Elsewhere in the state, the New York City Department of Parks and Recreation is reintroducing extirpated native species into local parks. Regardless of the species involved—big or small, reptile, amphibian, mammal, or bird—the same restoration principles and criteria apply.

Requirements for Success

In each successful case, restoration was both biologically feasible and socially acceptable. No species can succeed if the land will not support it; no restoration will work if the people on the land are against it. The more likely a species is to conflict with human activities, the greater the need for the informed consent of people who may not actually support restoration but will not oppose it. Although support for restoration can be widespread, one interest group or individual can stop a project. The biggest obstacles and challenges to government agencies are not in management or biological issues but in decision making. It is no coincidence that the common element in successful restoration efforts has been the species' compatibility with land use and broad public interests. Although the numbers and nature of advocates varied from case to case, there was no substantial opposition to previous successful restorations. In fact, management efforts for all restored species traditionally have been directed toward preventing their overharvest or overuse, not alleviating conflicts with humans. Even in the case of beaver *(Castor canadensis)* restoration, it was not until recent years, when pelt prices dropped substantially, resulting in a rapid population increase, that the species began to be considered more of a nuisance than an asset in some situations.

Most easy restorations are behind us. The larger remaining potential candidates, including elk, wolf, and cougar *(Felis concolor),* are much more problematic from the biological and social perspectives than were earlier successes. These restorations also will be more costly to undertake, more expensive to manage, less likely to succeed, and more subject to public debate than past projects. For these reasons, they must be held to higher standards of review and acceptance than the restoration of less demanding species. Wolf restoration is a case in point.

The DEC Position

Despite the recommendation in the 1992 recovery plan for the eastern timber wolf, DEC is not promoting timber wolf restoration to Adirondack Park because of limited resources, higher program priorities, and the questionable likelihood of success based on our current level of knowledge. Wolves will be no worse off tomorrow from a lack of immediate attention in New York than they were days, years, or decades ago. However, other species might suffer if existing resources are redirected toward wolf restoration. Nevertheless, DEC is willing to work with other parties interested in funding comprehensive feasibility studies to restore native species.

To date, DEC has not taken a formal position in support of or in opposition to wolf restoration and will not do so until more information is available on which to base a decision. Because DEC is the recognized management authority that must ultimately decide whether to permit the release of wolves if studies deem it feasible, it is important for DEC to remain objective and removed from the current process to the extent possible so as not to be perceived as a partner, opponent, or proponent in the current debate. For example, DEC declined an offer from Defenders of Wildlife (the not-for-profit environmental organization based in Washington, D.C., that has committed to funding feasibility studies in New York [see Schlickeisen, Chapter Seven, this volume]) to serve as a conduit for receipt and disbursement of funds to conduct the studies because it could compromise DEC's objectivity. In addition, to further preserve objectivity and impartiality, DEC turned down an invitation to participate as a voting member of the Citizen's Advisory Committee (CAC) (see Fascione and Kendrot, Chapter Six, this volume). However, DEC does participate as an advisor, together with the USFWS, at CAC meetings to answer questions about process and legalities. DEC also provided written recommendations in response to a request from Defenders for ideas and suggestions about developing feasibility studies.

To keep informed about the status, progress, and activities of the restoration initiative, DEC keeps abreast of day-to-day actions and developments on the wolf restoration question, including participating in CAC meetings, responding to questions and concerns from the public, and serving as a contact for various internal and external constituencies. If a proposal to restore wolves to the Adirondacks is presented to DEC for consideration, DEC will enlist the services of qualified staff, peers, and professionals who have not been actively involved in the issue on a daily basis to review it impartially.

DEC has indicated that any project proposal to restore wolves must first demonstrate that restoration would be ecologically feasible and desirable as well as socially acceptable. The department has not investigated any of these issues in depth and does not pretend to know the answers. It is likely that restoring wolves would be very difficult, particularly from the social and political perspective, and that social and political issues will drive the debate and ultimate decision about wolf restoration. Although public attitudes toward wolves have become more positive in recent decades and continue to change, there is still substantial opposition.

It is also likely that information in the initial feasibility studies will be insufficient to provide definitive answers to all the biological questions. As is generally the case, the most reliable way to answer many of the specific and detailed biological questions is by conducting on-site surveys and monitoring animals on the ground rather than embarking on costly and often speculative modeling, which time and fiscal constraints preclude from extensive exploration.

The Law

The decision to restore wolves to New York State rests with DEC, which is legally responsible for managing all wildlife of the State of New York. Section 11-0303 of the Environmental Conservation Law (ECL, New York State 1997) states,

> The general purpose of powers affecting fish and wildlife, granted to the Department by the Fish and Wildlife Law, is to vest in the Department . . . the efficient management of the fish and wildlife resources of the State.

It further states, "The Department is directed, in the exercise of the powers conferred upon it, to develop and carry out programs and procedures which will in its judgment . . . promote natural propagation and maintenance of desirable species in ecological balance."

Any attempt to restore wildlife species to the State of New York must be approved by DEC under Section 11-0507.3 of the ECL:

> No person shall willfully liberate within the State any wildlife except under permit from the Department. The Department may issue such permit in its discretion, fix the terms thereof and revoke it at pleasure.

In the specific case of wolves, even possession is strictly regulated under ECL Section 11-0511:

> No person shall, except under license or permit first obtained from the Department, possess, transport, or cause to be transported, imported or exported any live wolf, wolf dog, coyote, fox, skunk or raccoon.

DEC does not regulate for the sake of regulation. A permit and the associated review process pursuant to the State Environmental Quality Review Act are required to ensure that wildlife releases do not jeopardize the welfare of the public or the environment. The consequences of unregulated releases or escapes, such as starlings *(Sturnus vulgaris)* and house sparrows *(Passer domesticus)*, have taught us to be careful when considering the release of wildlife species. Furthermore, a Draft Environmental Impact Statement that systematically addresses significant environmental impacts, explores various alternatives including a no-action alternative, recommends mitigative actions if appropriate, and incorporates public viewpoints and concerns also must be developed. Ultimately, public support, or the lack thereof, will be assessed by DEC through some process of public participation.

The Issues

Some of the issues that must be addressed to DEC's satisfaction before it approves a proposal to restore wolves include the following.

Historical Status

By definition, restorations involve native species. Although it seems reasonable to assume that wolves were once present, their historical status must be investigated thoroughly. The Adirondack Park State Land Master Plan (Adirondack Park Agency 1979) states, "There will be no intentional introduction in wilderness areas of species of flora or fauna that are not historically associated with the Adirondack environment." Currently, there is scant evidence in the literature that wolves existed in the Adirondacks or New York. One also cannot overlook the fact that the eastern coyote *(Canis latrans),* which inhabits the Adirondack Park, is larger than its western cousin and exhibits many wolflike characteristics. Is it a wolf–coyote hybrid and, if so, to what extent may we already have some form of wolf in the Adirondacks? Genetic studies must be completed to determine the makeup of our native wild candid.

Biological Feasibility

Can the Adirondacks sustain a wolf population over the long term without constant and intensive intervention by DEC staff? Is there sufficient land, prey, and isolation from human activity to give a wolf population a reasonable chance of success? Can wolves survive at the current human density? Is the Adirondack Park large enough, or should we instead take a larger look at northern New York and perhaps southern Canada and Vermont? Is the Adirondack Park conducive to wolves remaining within the bounds of the park, or will they emigrate to more suitable areas outside the park where private land is predominant and prey resources more abundant? Given the geographic isolation of the Adirondacks from the nearest naturally occurring wolf populations, there is little chance of new recruitment. Will an isolated wolf population be able to survive on its own, or will restoration result in an island population for which long-term sustainability is tenuous? In the absence of outside recruitment, to what extent will inbreeding affect a geographically isolated population? Other biological concerns revolve around coyote–wolf interactions. On one hand, some will rejoice at the distinct possibility that wolves will outcompete, drive out, or kill coyotes within their territory. On the other hand, this intense species competition is seen as a liability. Perhaps more problematic is the potential for hybridization between coyotes and wolves, and it is unknown whether a hybrid is in the best interest of wildlife species management.

Ecological Consequences

What changes, good or bad, are likely to occur in the existing wildlife populations of northern New York if wolves are returned? Species of interest include white-tailed deer *(Odocoileus virginianus)*, beaver, coyote, varying hare *(Lepus americanus)*, and moose. There are serious questions about whether the current prey base is adequate for long-term sustainability. The Adirondacks support a population of deer that ranges from two or three per square mile to 30 deer per square mile (Sage, 1997, personal communication; Sage, Chapter Four, this volume). By some estimates, it takes 10 deer per square mile to support wolves, which is the Adirondack average. However, in 19 of the past 30 years, deer populations have been below 10 per square mile (below 5 per square mile in 6 of those 19 years) and in only 11 of the past 30 years above 10 per square mile in (Sage 1998; Sage, Chapter Four, this volume). Is this adequate, or would there be significant impacts on other wild and domestic species? Also interesting is the fact that there are more deer on private, managed land within the Adirondacks than on public land and that there are more deer on private land outside the Adirondack Park than on private land within the park. Given that wolves are a prey-based species, it seems logical to assume that they will be attracted to private lands more often than public lands, which increases the potential for human–wolf conflicts. One also has to question the potential impact on the slowly increasing moose population and ask whether the presence of wolves will significantly slow or reverse the recovery of the small moose population.

Consequences to People

What would be the positive and negative consequences of a wolf population to the people of the State of New York, especially northern New Yorkers? Would they welcome or at least be willing to live with those consequences? Would DEC have the authority to deal with conflicts? Some of the issues that might arise include the positive and negative effects of wolves on the local economy, deer hunting, domestic animals, land use regulation, and hunting and trapping regulations. What attributes and liabilities are associated with wolf restoration? Property rights, aversion to government intervention, and home rule are values that Adirondack residents cherish and defend. Wolf restoration is viewed as a threat by many, not necessarily to other wildlife or species but to a way of life that many believe will diminish their property rights and increase government interference in their private lives. There is also the very real possibility that wolves, like the lynx, will wander great distances, and potential impacts on neighboring Canada and Vermont must be considered during planning.

Social Acceptability

DEC will not support wolf restoration without the informed consent of the people of the state, particularly those in the affected area. There are many stakeholders in the wolf debate on both sides of the issue. Any one of them, if sufficiently motivated, can prevent wolf restoration. But whom do you ask: local residents most likely to be affected by wolf restoration, New York State residents (because all, at least in theory, own a piece of the Adirondack Park), or the public at large (because the wolf is a public resource)? Although some studies suggest that there is strong individual support for wolf restoration, numerous organizations and local government entities have publicly voiced strong opposition and many environmental organizations have remained silent on the issue. Two counties in the Adirondacks have passed local laws prohibiting the transportation and liberation of live dangerous animals, including wolves. Although the local laws are legally viewed as being beyond the counties' authority to enact, they nevertheless make a strong statement that must be considered. The question remains, however, as to whose opinions count and how one weighs individual preferences when there are such strong opposing views.

Costs to DEC

It is inevitable that DEC will become actively involved in the wolf restoration debate, given its regulatory and management responsibilities. Regardless of who conducts the study or submits an application, DEC will be involved in reviewing the feasibility studies for completeness and accuracy to determine whether additional information is needed and whether the findings and recommendations are sound and addressed accurately. If restoration proves feasible and is implemented, it must be determined who will carry out the implementation and at what cost, not only out-of-pocket expenses but costs to and impacts on other programs and projects. If it is implemented, DEC will be responsible for managing wolves as long as they persist. There is concern about potential management costs for monitoring, mitigation, and enforcement. What are the short- and long-term costs to DEC of managing a wolf population? Where would the funds come from to pay for this management? How would it affect existing programs and priorities? There may also be legal obstacles, lawsuits, or prohibitions preventing DEC from becoming involved in wolf restoration, all of which may involve a tremendous amount of time and cost.

Conclusion

Any species restoration, and wolf restoration in particular, must be a well-deliberated process, not to be rushed or entered into lightly, no matter what the temptation or popularity. We do not want to turn the wolf restoration issue into one in which wolves in New York represent hate, mistrust, and government or environmentalist intervention and land use regulation. Given the likelihood of sharply conflicting, deeply held views toward wolf restoration, it is important that all parties proceed slowly, giving everyone the opportunity to be heard while considering their values and concerns. It is important that wolf restoration studies and decisions proceed in a scientifically based, reasonable, and responsible manner.

Part II

Voice: The Interests
and Concerns of Stakeholders

In this part we bring together the perspectives of four stakeholders in the Adirondack question and a sociological analysis of public survey data on restoration efforts generally.

In Chapter Four, "Wolves in the Adirondacks? Perspectives from the Heart of the Adirondack Park," Richard Sage speaks not only as an Adirondack resident but also as chair of the Town of Newcomb Planning Board and a professional forester and wildlife researcher who is associate director of the Adirondack Ecological Center.

Speaking as a forester and wildlife researcher, Sage presents data on deer density in the park. Studies of the central Adirondacks show that for 19 of the last 30 years, deer density has been below the estimated 10 deer per square mile minimally necessary to support wolves. Estimates from 1999 place the population at 6 to 8 deer per square mile. According to Sage, deer densities in the central area of the park are influenced primarily by severe winter weather, poor habitat quality, and wild canid predation. The unpredictable nature of winter weather, combined with the declining quality of habitat on Forest Preserve lands, is unlikely to improve white-tail population numbers in the decades ahead.

Extrapolating from the reintroduction of the Canada lynx into the park in the late 1980s, Sage observes that although much of the land in the park is preserved under the "forever wild" designation, the park is best regarded as a compromised wilderness from the perspective of wildlife habitat. Roadkills within the park boundaries were the number-one source of documented mortality of introduced lynx.

As an Adirondack resident, Sage reports on the decline in the economic base of the Town of Newcomb since the 1970s. With the decline of the timber industry and the closure of a titanium mine, the population has dropped from 1,500 to 544. Young people have left for jobs elsewhere, the population has aged, local businesses have closed, and protective services have become strapped for volunteers.

Speaking as the 18-year chair of the Newcomb Planning Board, Sage recounts the adversarial relationship that has developed between the town and the Adirondack Park Agency on land use issues. Of the town's 218 acres, the local community has primary say over development and land use on only 4 square miles. With no gas station or motel in the town and almost no land available to build tourist amenities, the town is unlikely to capitalize on any increased tourism that might be associated with wolves in the Adirondacks.

Given these circumstances and the high level of biodiversity already existing in the park, Sage, like many other residents of the park, sees little compelling reason to restore wolves to the Adirondacks.

Eli Thomas is a member of the Wolf Clan of the Onondaga Nation. The Onondaga, like the Mohawk, Oneida, and Cayuga, inhabited the Adirondack region long before settlement by whites. In Chapter Five, "The Two-Leggeds and the Four-Leggeds," Thomas offers a meditation on the wolf, its power, and its relationship with humans.

Drawing on the diaries and town meeting records of white settlers to Manlius, New York, Thomas tells of the fear that resulted in bounties for panthers and wolves. In the Onondaga oral tradition, however, the Wolf Clan consists of the brothers and sisters of the wolf. Drawing on this oral tradition, Thomas calls for trust between the human players in the Adirondack debate as a necessary prelude to a time when wolves and humans may once again walk the same path.

This part also includes two chapters from Defenders of Wildlife, the national environmental organization that introduced the restoration proposal and is paying for the initial biological and socioeconomic feasibility studies.

In Chapter Six, "Facilitating Citizen Participation in Adirondack Wolf Recovery," Nina Fascione, director of carnivore conservation, and Stephen Kendrot, former northeastern field representative for Defenders of Wildlife, first explain Defenders' role in prior wolf restoration efforts. They point out the geographic barriers to natural wolf recovery in the Adirondacks and report on a 1996 Responsive Management poll indicating 80% approval among New Yorkers and 76% approval among Adirondackers of wolf restoration to the area. They devote the bulk of their chapter to describing the formation and work of the independent Citizen's Advisory Committee (CAC) that has overseen the process of soliciting and vetting proposals for the feasibility studies on wolf reintroduction in the Adirondacks.

Although trust between the diverse members of the CAC has been hard won, state and federal agencies and the media have praised the process as a model for community participation in wildlife restoration efforts. Having participated as nonvoting members and observed the committee's function, Fascione and Kendrot observe that the group dynamic ensured that qualified researchers conducted the studies. Because the public plays such an important role in modern conservation, the authors say, the CAC process proved to be an important mechanism for including public deliberation and participation in decision making. The committee process has been invaluable not only in developing a framework for how varied groups can discuss contentious issues but in enabling Defenders to better understand northeastern stakeholder concerns and issues about wolf recovery and more effectively merge advocacy with public involvement.

In Chapter Seven, "Overcoming Cultural Barriers to Wolf Reintroduction," Rodger Schlickeisen, president of Defenders of Wildlife, underscores one of the major points of agreement in the volume, namely that wolf reintroduction can occur only if the undertaking is scientifically and financially feasible and if the public wants it to happen. In a brief review of previous reintroduction efforts in Yellowstone and central Idaho, Schlickeisen concludes that whereas the first prerequisite for reintroduction is easily satisfied, the second is more problematic.

Citing centuries-old hostility toward wolves, Schlickeisen observes that continuing antipathy is based on a number of misperceptions and exaggerations. First, although there are no known instances of wolves attacking humans in the United States, people continue to believe that wolves pose a threat to human safety. Second, says Schlickeisen, a number of fears, though not baseless, are exaggerated, including the fear that wolves will kill pets and livestock and decimate the wild ungulate populations. Third, in many cases, opponents of wolf reintroduction have exaggerated the likelihood that wolf reintroduction would be accompanied by land use restrictions. These fears are manifested in illegal killing of wolves and legislative attempts to block restoration efforts. Schlickeisen acknowledges that opponents of reintroduction have reasonable concerns and that many environmental activists are dismissive of local issues.

To address these fears and concerns, Defenders has developed a number of outreach and education programs, including the livestock compensation fund, and has promoted active involvement of local community representatives in developing reintroduction proposals.

Schlickeisen concludes that prospects for success of a reintroduction effort increase significantly if wolf advocates show respect for the rural residents' cultural perspectives and respond with education to counter negative mythology, financial assistance to compensate for predation on livestock, and willingness to limit regulatory requirements to what is necessary. Success in these efforts can facilitate reintroduction and reduce illegal wolf killing after reintroduction. Finally, wolf reintroduction is also a political decision, subject to the vicissitudes of the changing political landscape. Wolf advocates need to respond with vigilance and persistence, he says, and demonstrate their own political power and strategic flexibility to overcome the hurdles.

This part is rounded out by Chapter Eight, "In Wolves' Clothing: Restoration and the Challenge to Stewardship." Jan Dizard, Charles Hamilton Houston Professor of American Culture at Amherst College, uses public survey data as a basis for reflection on public perception of restoration efforts generally. He describes two narratives—one of loss, one of recovery—that have dominated our thinking about nature and shaped our contradictory and often embattled attempts to define an ethically and materially sustainable relationship with the natural world. In their extreme versions, these two narratives are diametrically opposed. The narrative of loss becomes a wholesale condemnation of modern society and an evocation of a fast-approaching apocalypse. At the other extreme, the narrative of recovery can provide a fig leaf of respectability for the so-called Wise Use movement and others who reject almost all environmental regulation and restraint on our exploitation of nature. The narrative of recovery also underpins the views of those few scholars who remain convinced that nature's bounty is unlimited, capable of absorbing very large increases in human population.

Fortunately, he maintains, we do not have to be captive of either narrative—at least not yet. Reckoning with our many losses, we have begun to develop and refine ways to reclaim and restore habitats and reintroduce wildlife species to habitats from which they have been driven. Many promising efforts are under way to boost the recuperative capacities of nature, including restoration efforts with prairie, wetlands, and species such as the California condor, peregrine falcon, and wolf.

Although this resurgence is encouraging, he says, no one should imagine that the cumulative effects of thousands of years of human appropriation of nature can be undone. Many life forms have gone extinct by virtue of our adaptive success, and it is clear that even with the best intentions and an unimaginably concerted effort, many more species will disappear by our hand. This will continue to fuel the sense of loss and, for some, give urgency to the desire for recovery. Herein lies the danger. If we are too indiscriminate in our efforts, we will exhaust ourselves by trying to do the impossible; if we are too narrowly selective, if we focus our efforts only on a few charismatic species and inspiring habitats, we run the risk of making things far worse than they might otherwise be.

Dizard closely examines opinion polls on the public's motivation for natural restoration efforts and finds some dangerous misconceptions about our relationship to the natural world. The dominant one is that if we can set aside habitat and keep our hands off, everything will work out for the best; wild animals can take care of themselves if we leave them and their habitat alone. This view of nature and our relationship to it casts a thick veil of suspicion over all efforts to manage natural resources except those that can be cloaked in restorationist garb. Benign neglect replaces stewardship and relieves us of all the sticky moral problems attendant on active management of natural resources. Like curators of a museum, we are responsible for ensuring that the collection is not misused or contaminated by outside influences (i.e., people).

To the contrary, Dizard argues, the real challenge of stewardship is to accept that reintroducing species is only the beginning of long-term responsibility for informed and ongoing management of habitats and wildlife. Thus, those who would promote wildlife must make clear that creating good habitat for wildlife and reintroducing species does not end our responsibility. In fact, it increases our responsibility and increases the need for management. Setting aside habitat will not produce harmonious nature any more than introducing wolves will restore balance to an area that wasn't balanced to begin with. The return of the wolf promises to increase our love of the wild. If the wolf also helps us recapture a sense of responsibility and stewardship, we will owe it a great deal. If we duck our responsibility, we will have succeeded only in victimizing the wolf yet again, this time with good intentions.

Chapter Four

Wolves in the Adirondacks? Perspectives from the Heart of the Adirondack Park

Richard W. Sage, Jr.

I have been very fortunate to have lived and worked in the small hamlet of Newcomb, located in the very heart of the Adirondack Park, for the past 31 years. As a professional forester, wildlife researcher, and current associate director of the Adirondack Ecological Center (AEC, a research station operated by the State University of New York College of Environmental Science and Forestry), I have had the unique opportunity to experience and learn about the Adirondack ecosystem, its forests and wildlife, and its people and politics. My comments about the potential of wolf restoration to the Adirondack Park will first address some things we do and do not know about the science of wolves in the park. Second, I want to make some observations as a long-term resident of the park and then address the question of wolf restoration from my viewpoint as chair of the Town of Newcomb Planning Board.

We know a great deal about the Adirondack ecosystem, and some of what we know has direct implications for the question at hand. Is there a place for wolves in the park?

At least four preliminary assessments of the potential of the Adirondack region to harbor wolves have been made to date (Henshaw 1982; Hosack 1996; Hodgson 1997; Mladenoff and Sickley 1998). These appraisals were based largely on human population density, road access, and abundance of prey populations. Although these factors have been used to identify appropriate areas for wolf reintroduction, a much more comprehensive approach is necessary to evaluate a region's potential. Such investigations must include human or social issues as well as biological considerations. Currently, a comprehensive study funded by Defenders of Wildlife is under way. Unfortunately, many people have already made up their minds on both sides of the issue. As both a scientist and an interested resident of the park, I am concerned about how effective the findings of this study will be in influencing the final decision.

The Adirondack Ecosystem: My Perspective as a Forester and Wildlife Researcher

Research at the AEC provides some critical information relating directly to the wolf reintroduction issue. These data were gathered in the Adirondack Park, so it is not necessary to draw inferences based solely on data from distant, very different ecosystems. Experiences in Yellowstone, British Columbia, or Minnesota certainly are of interest but do not realistically represent conditions in the Adirondacks. Likewise, the absence of permanent residents, human communities, and privately owned land makes comparison of Yellowstone with the Adirondacks inappropriate.

White-tailed deer undoubtedly will be the primary prey species of any future Adirondack wolf population. Minimal densities of deer necessary to support wolves have been established at about 10 deer per square mile. Our data clearly show that during the past 30 years deer densities in the central Adirondacks have been below this level for 19 years and at or above this minimum for only 11 years. In 6 of the 19 low years, deer densities were less than 5 deer per square mile (Demers 1998). Currently, deer densities in the New-comb area are 6 to 8 deer per square mile (Nesslage 1999), and the New York State Department of Environmental Conservation estimates deer densities at only 5 to 6 deer per square mile within the Adirondack Park (Armstrong 1998).

Our research has shown that deer densities in the central area of the park are influenced primarily by severe winter weather, poor habitat quality, and wild canid predation (Underwood 1986, 1990). The unpredictable nature of winter weather, combined with the declining quality of habitat on Forest Preserve lands, probably will not improve white-tail population numbers in the decades ahead.

The impact of wolf predation on the prey base has been an issue of great debate. Our population models clearly indicate a significant negative influence on local deer populations attributable to wild canid predation (Brundige 1993). This impact is most limiting when deer densities are reduced by severe winter weather. Predation has prolonged the recovery period by as much as four years (Sage 1998). Our studies have clearly shown that the eastern coyote (as it is currently called) is an effective predator on both fawns and adults throughout the year but particularly in spring and winter. As much as 80% of the winter diet is composed of deer, resulting from predation and scavenging of winter-killed animals. In spring, fawns are the primary age class affected (Chambers 1987; Brundige 1993). In winter, all age and sex classes are vulnerable because of the restricted movement conditions imposed by deep snow and the weakened physical condition of the animals.

Studies of the current Adirondack canid conducted through our research center show behavioral traits characteristic of wolves: significant use of large prey, hunting in small groups, and seasonal shifts in territory associated with prey movements (Brundige 1993). Morphologic measurements of canid carcasses from across the region, including weight, length, and skull measurements, place this animal between coyote and wolf. Evidence gathered from carcasses collected and submitted for DNA analysis strongly suggests a close relationship between the Adirondack canid and the Algonquin wolf. Furthermore, both of these animals appear to be most similar to the red wolf rather than the gray wolf (Chambers, personal communication, 1999).[1] As more information is gathered, these genetic links can muddy the waters relating to wolf reintroduction into the Adirondack Park even further.

Research at the AEC has identified other ecological relationships that bear on the question of wolves in the park. The Adirondacks are not a highly productive ecosystem. Generally poor soil fertility, combined with a short growing season and harsh winter weather, result in lower productivity than typically associated with our native animal species. This is manifested in lower birth rates, reduced survival of young animals, increased age at maturity, large home ranges, and low population densities (Gustafson 1984; Fox 1990; Costello 1992; McNulty 1997).

We have also learned a great deal about the limits of the Adirondack wilderness from our experience with the reintroduction of the Canada lynx in the late 1980s. From the outset it was clear that the influence of people and their associated development is pervasive throughout the park. Roadkills were the number-one source of documented lynx mortality within the park boundaries. Animals routinely visited dumpsters, parking garages, and roadside environs. Their extensive wanderings clearly indicated that they had little trouble traversing the breadth of the park and in many cases exiting the park and the State of New York altogether (Gustafson and Brocke 1990, 1998). The presence of 130,000 permanent and 110,000 to 200,000 seasonal residents, together with 6 million visitors annually, and the infrastructure to support them, cannot be overlooked as we address the question of wolf reintroduction. As much as we might like to think of the Adirondacks as a vast wilderness, in reality it is a compromised wilderness at best.

Human Communities in the Adirondacks: My Perspective as an Adirondack Resident

Let me switch to my Adirondack resident hat, in particular my Town of Newcomb hat. First, let me tell you how to get to Newcomb. You go out the door of the American Museum of Natural History at 77th Street in New York City, turn right and walk down to the river, put your canoe in the water, and paddle upstream for about 20 days. When you get to a spot where the river is no wider than your canoe, you have arrived in Newcomb, the geographic center of the Adirondack Park. It is a magnificent setting, looking north over a panoramic view of the high peaks, including Mount Marcy, the highest point in New York State and the uppermost reaches of the Hudson River drainage. Unending hardwood and conifer forests and myriad lakes, ponds, and miles of wild, scenic, and recreational rivers provide habitat for a variety of wildlife including moose, black bear, fisher, marten, and otter.

When I first came to Newcomb in 1964 I was immediately impressed by this setting and the people of this proud and prosperous out-of-the-way community nestled in the shadow of Santanoni Mountain. At this time, the population of Newcomb was around 1,000, growing to nearly 1,500 by the early 1970s. Our school population was more than 400 students (K–12). The titanium mine was in full operation and the regional timber industry was strong. The area deer population was at an all-time high, and hunting and fishing enthusiasts were important contributors to the local economy. Newcomb's economic base has always been tied directly to the management and use of its natural resources. Times were good in Newcomb.

Today, our population is just over 500. Our school district is the smallest in New York State, with only 64 students in grades K–12. The mine closed in 1987 despite the fact that there are 50 million tons of ore still in the ground. The timber industry has had its ups and downs, and the deer population, after falling to an unprecedented low, has never recovered. Times have changed for Newcomb (Town of Newcomb 1990).

As you might expect, these events weigh heavily on the local residents. We are fighting for our survival and continued existence as a community. Our young people must leave to make a future for themselves. The resident population is aging. Local businesses are closing, and protective services are strapped for volunteers. Collectively, these factors are the formula for extinction.

When you put our rather ominous future together with the fact that I already live in a high-quality environment, with a high level of biological diversity, and have almost daily opportunities to experience unique wildlife and breathtaking scenic vistas, it is not surprising that I (and many other Newcomb residents) do not feel compelled to restore wolves to the Adirondack Park. Instead, I feel compelled to see to it that my community survives in the park, that my son and daughters and others also have the opportunity to experience what I have enjoyed for the past 31 years, living and working in the heart of the Adirondack Park. In order for that to happen there must continue to be a Newcomb Central School and a proud and prosperous community nestled in the shadow of Santanoni Mountain. This is by far my biggest concern, and I am certain it is a very real concern to the 544 other people who live in Newcomb.

The Adirondack Park Agency: My Perspective as Chair of the Newcomb Planning Board

I have served as chair of the Town of Newcomb Planning Board for the past 18 years. To be honest, it is a thankless, time-consuming job without many rewards. During my tenure, the town went through the very long and arduous process of developing a local land use and development plan (zoning), which included subdivision regulations and on-site sewage disposal guidelines. All such plans developed by communities within the Adirondack Park must be consistent with the rules and regulations set forth in the Adirondack Park Agency Act. This legislation, passed in 1972, also created the Adirondack Park Agency (APA) as the regulatory agency of the state, responsible for overseeing land use and development on all private lands within the Adirondack Park.

As you might expect, state regulation of the use of private lands in the park was not well received by local residents. Resentment of the agency and what it represents has been the basis for a continuing controversy since its creation. There is a strong feeling among private landowners that the agency far exceeds its legislative authority in pursuing an antidevelopment, preservationist agenda promoted by agency staff and commissioners who do not represent the interests of park residents.

Our experience with the agency in attempting to get the Newcomb land use plan through the APA approval process supports this contention. This process lasted more than two years and pitted the Town of Newcomb against the agency at nearly every step of the way. As submitted, our plan included all the rules and regulations required in the Park Agency Act and was even more stringent on some issues. An agency staff member worked closely with our planning board throughout the development process, reviewing every section along the way. However, when the plan was completed and submitted for approval, the agency decided that the "rules" (as they would like to see them, not as the law specified) had changed and that our plan needed sweeping changes. Coincidentally, as our plan went before the agency, a state-sponsored study commission report on the Adirondacks in the twenty-first century had been released calling for more stringent controls in the park. These were the same conditions that the agency tried to impose on us, even though no legislative action has ever been taken on the report's recommendations. Our plan eventually was approved by the APA, basically as written, with minimal changes. However, the overall experience could hardly be called a positive example of local and state government cooperation.

This history of confrontation, perceived overstepping of authority, indifference to residents' concerns, overt alignment with environmentalists, and outright arrogance on the part of the APA has in part created an adversarial relationship and a feeling of mistrust among many park residents, particularly when faced with new initiatives affecting the Adirondacks. It is under this cloud of suspicion that many park residents view the question of wolf reintroduction.

Despite all our efforts, what has our approved local land use plan gained us? There are 218 square miles of land in the Township of Newcomb. The state owns half of this land outright as part of the Forest Preserve. No development of any kind is permitted on these lands. All remaining private land is subject to the controls of the APA, acting as the state's zoning authority within the park. We, the local community, have the primary say over development and land use on only 4 square miles (the hamlet area) of our town. Even this land is subject to APA authority if it includes wetlands, shorefront property, or scenic vistas. The question becomes, Where is Newcomb going to build its future? We find ourselves land poor in the midst of plenty.

Here is another bone of contention that places many Adirondack residents on opposite sides of the fence with issues such as wolf reintroduction. Employees of New York State and people associated with private environmental groups are working very diligently to acquire more private land within the park to be added to the Forest Preserve as "forever wild" land. This land is being acquired with money from bond acts passed by the votes primarily of New Yorkers living outside the park. Nearly 200,000 acres of land have been added to the 2.4 million already owned by the state within the past few years. A continuing argument for expanding the Forest Preserve is as a place for moose, panthers, and wolves. Many Adirondack residents see this state land acquisition policy as a continual erosion of the natural resource base on which our local economies depend. The wolf reintroduction issue is seen by many residents as a front for further land acquisition and a potential Adirondack "spotted owl," leading to more regulation of private land within the park.

In place of our well-paying, skilled, year-round jobs in mining and the timber industry, tourism is offered as the salvation of our community. Give up on resource management and use as an economic base, we are told. For tourism to work, you must have an infrastructure in place to capture money from visitors, such as restaurants, motels, shops, services, and theme parks. We have none of this now and almost no land available on which to build it. Currently, we don't even have a gas station in town. Furthermore, a survey of town residents indicates that many don't want this kind of environment (Town of Newcomb 1990).

Our planning board has looked hard at outdoor recreation–oriented tourism as one way to bolster Newcomb's economy. There appears to be a high level of compatibility between these activities and the Newcomb landscape, but the lack of an established service infrastructure severely limits our options. The inability of the New York State Department of Environmental

Conservation to complete its mandate to develop unit land management plans (now nearly three decades behind schedule) for all state-owned lands in the park, including those in the Town of Newcomb, also stands as a major obstacle in our efforts to move ahead with any outdoor recreation planning efforts.

Reestablishing the wolf in the Adirondack Park has been cited by some as a potential boon to area tourism. The credibility of this claim is difficult to evaluate at this time, but one thing is certain: Newcomb is not in a position to capitalize on the presence of wolves. Actually, our appraisal of the situation is quite to the contrary. Area deer populations already are low for a variety of reasons discussed earlier. The presence of wolves in the park (which will find private managed timberlands more to their liking because of higher prey densities than those on state Forest Preserve land) could depress numbers even further. Certainly, wolves will not have a positive influence on deer numbers. Even if wolves have minimal impact on deer numbers, the perception of most deer hunters is that hunting success will decline. Whatever the case may be, reality or perception, it could have a direct effect on our local economy. Newcomb has a long history as a popular deer- and bear-hunting area. Local timber companies lease approximately 95% of their lands to hunting and fishing clubs. These leases are a significant source of revenue to private landowners in the Newcomb area. Hunting and fishing camps also are a part of the tax base in Newcomb, and the money spent on hunting and fishing is an important source of local revenue. We are not inclined to trade an established revenue-generating tradition for an incompatible and unproven venture into wolf-based ecotourism.

Conclusion

In summary, I think it is critical that we recognize and incorporate the extensive Adirondack-based ecological research information into our evaluation of the question of wolves in the Adirondack Park. Successful experiences in very different ecosystems by no means ensure success in the Adirondacks. We must be objective, thorough, and professional in assessing the potential for wolves in the park.

I have also tried to express some perspectives from which many local residents of the park view the question of wolf reintroduction to the Adirondacks in hope that it might help explain why not everyone is enthusiastic about the idea. Without question, some park residents favor wolf reintroduction, just as some vehemently oppose the idea. I think many of those expressing opposition may not necessarily be against the wolf itself but against what the wolf could bring in terms of further land use restrictions on private lands and increased acquisition of state land. The events of the past three decades lend credibility to their concerns.

When it addressed critical issues relating to wolf reintroduction to Yellowstone Park, the U.S. Fish and Wildlife Service made an observation that is probably one of the few things also applicable to the Adirondacks: "Local acceptance of the wolf recovery program is desirable, if not necessary. Members of the public that live, work, and engage in recreational activities in the wolf's habitat can make the difference whether a population exists or not" (Yellowstone National Park 1990, 24).

In light of this statement, I find it interesting that the residents of the Adirondack Park must struggle to be heard and feel that their viewpoints are not likely to weigh heavily in the decision-making process. We should never lose sight of the fact that if wolves are reestablished in the park, these animals will live in our backyards.

As a result of more than 30 years of living in the park I am convinced that outsiders are incapable of seeing things the way we do and insiders are incapable of seeing things as they do. This is not said in criticism of anyone on either side of the Blue Line, but I believe that this explains why our viewpoints nearly always differ. Our perspectives and priorities will never be the same. For this reason, we must be given a fair hearing.

Endnote

1. R. E. Chambers is a professor at State University of New York College of Environmental Science and Forestry, Syracuse.

Chapter Five

The Two-Leggeds and the Four-Leggeds

Eli Thomas

My name is Eli Thomas, and as I write this I am very happy and am having an enjoyable time. I appreciate having the opportunity to give you my message and to share my thoughts of those who came before me: my mother and my father, and my uncles and my aunts and my grandmother and all as far as my memory travels back.

I probably do not speak only for myself but for many people who will agree that the gratitude we send and the thankfulness we feel from youth is little to ask in relation to the gifts we experience daily. The beauty of a waterfall, the colors of a rainbow, or the peak of a mountain are all things we should admire. We should pause for a moment and sincerely give thanks from our hearts. As we express our thankfulness, we need to think about the water we drink as it falls from the sky and eventually surfaces from the earth. The cycle of nature cleanses and freshens the water so that it quenches our thirst and enables us to grow strong. The trees, the four-leggeds, six-leggeds, the things of the water, the things with wings, that which is beneath earth, are all part of us. We mention them in gratitude today.

The natural world has traveled with us; we have traveled together. As a person from the longhouse tradition, I see us as part of an extended family. In the Mohawk tradition, our ancestors have a path that goes across the sky, and as we journey we may sometimes go back to our relatives and forward to our future generations. If we express a good mind and a good heart now, the seventh generation to come will experience a healthy earth and as enjoyable a journey as we do.

I am a full-blooded Onandaga. My blood and spirit are indigenous with the land, the language, and those who came before me. Onandaga traditions are strong and have not changed. I am from the Wolf Clan; my family is wolf. We look at the four-leggeds as relatives and teachers. To help you understand the indigenous perspective, I will go back before 1700 to a place we now call Manlius, New York. Manlius is about 30 miles south of the Adirondack Park and is part of the Erie Canal system. My ancestors, Onandaga people from the

47

Wolf Clan, lived 4 miles south of Manlius. They lived in 175 longhouses averaging 100 feet long. Longhouses are estimated to have housed one person to
a foot, so it would be around 17,500 people living in one village, one family.
Further east would be people from the Deer Clan, Onandaga. A little bit further east, people of the Turtle Clan, Onandaga. And down around Onandaga
Lake, people from the Eel Clan, Onandaga. Further south near Cortland, New
York, people from the Beaver Clan, Onandaga, and so on for nine different
families in various areas. And then our little brothers, the Oneida Nation, had
only three families, but they were set out in the same pattern; if you could be
up in the eyes of an eagle and look at those families, it would be in the same
pattern with the Oneidas and the Cayuga Nation. That was before the roads,
electricity, and fences. The Senecas had the same. They had nine families. And
the Mohawks had four families. The Mohawks are the indigenous people who
seem to be a little bit more affected by the idea of wolf restoration in the
Adirondacks. I have received in an indirect way from the Mohawks and the
Onandagas approval for how I conduct myself.

Before the building of the Erie Canal, immigrants settled in the area
around Manlius. They came by foot or by oxen, carrying whatever they could.
They were alien to the land. I researched the diaries these people kept to discover what they said and felt. The record of one town meeting included discussion of a bounty for panthers and wolves. Apparently panthers and wolves
were abundant, and these people were afraid to sleep at night for fear of the
wolf's howl. The diaries said that the people stood with clubs behind their
backs, and when the wolves came down the path, they used these clubs to kill
them. Our oral tradition tells us that before electricity and roads, our people,
the Wolf Clan of the Onandaga, as brothers and sisters of the wolf, could walk
down a path past wolves without fear. We had respect, we were related, and the
wolves and my ancestors shared trust. When the wolf walked down the path,
we would step aside and let it pass. We respect the natural world and all living
things. Respect is essential to understanding. We understood. The settlers did
not understand, for they were fearful in their hearts.

Fear breeds negative thoughts. Fear leads people to believe they are surrounded by enemies. If we feel fear and perceive enemies, perhaps we should
face our fear by educating ourselves. We can do so by talking and sharing a fire
together. There is a language between my people and the wolf. As in sharing
the fire, we share a place in this world together. Our eyes reflect the campfire
in the darkness.

We, as people, can be transformed. When we encounter hardships, the
watchful wolf helps us to regain our strength. They lick our wounds and, if we
are lost, tell us how to get back to our relations. They provide us with sustenance whenever necessary. Their howling provides comfort to us. Rationally,
we trust our relatives and our teachers. That trust is a precious gift. We can't

touch or see it, but in a way that gift is symbolic of the wolf and my family. We want to care for things around us. That is what the creator intended: a sharing experience, not separate things. All things are related and connected.

Having the tahionni (the wolf), the relative and teacher, in the Adirondacks can be done. I have heard from the oral tradition of my people that wolves and people can travel together down the path. Once again, we can share the gifts of the natural world as creation continues to grow strong.

Let us agree that the eagle is resting on top of the tree of peace. As long as the eagle is present there will be peace on earth for anyone who finds shelter under its branches. A wolf may appear to be hiding, but it is peacefully resting under those branches too. That's the way it is.

Chapter Six

Facilitating Citizen Participation in Adirondack Wolf Recovery

Nina Fascione and Stephen R. Kendrot

From a distance it appeared as though many audience members in the State University of New York's Plattsburgh campus auditorium had wolf images on their t-shirts. For a few moments I was under the illusion that the more than 400 people who had gathered on this early spring night in 1997 for a 2-hour panel debate were primarily sympathetic to the topic I was there to discuss: potential wolf restoration in the Adirondack Park of upstate New York. The Plattsburgh debate, organized by interested people from the college and local media, was one of the first large public forums to discuss the issue. As I made my way to the speaker's table in the front of the room and got closer to the audience I realized that although there were indeed wolves on many of the shirts, the majority were pictured in the crosshairs of a rifle scope. I suddenly realized that I, as a representative from Defenders of Wildlife, was entering a much less hospitable environment than I had anticipated.

Wolves, the top predator of the Northeast's Great North Woods for centuries, had been missing from this ecosystem since the late 1800s because of years of human persecution. Defenders of Wildlife, a national conservation organization headquartered in Washington, D.C., had recently begun to investigate whether the Adirondacks could still hold wolves after 100 years of growth and change. Several years earlier, the U.S. Fish and Wildlife Service (USFWS) had identified the Adirondacks as one of three areas in the Northeast with potential as wolf restoration habitat in their 1992 revised *Recovery Plan for the Eastern Timber Wolf,* with two areas in New England as the other sites.

However, the geographic scope of the recovery plan was huge, designed to recover wolf populations from Minnesota to Maine and all northern states in between. According to the plan, complete recovery of the eastern timber wolf would be satisfied by establishing two wolf populations: one in Minnesota, where wolves were never fully extirpated, and a second population outside Minnesota. This second population, if established within 100 miles of Minnesota's wolves, should contain at least 100 individuals. A second population farther than 100 miles from Minnesota must contain more than 200 wolves to meet recovery criteria (U.S. Fish and Wildlife Service 1992). With wolves dispersing from Canada and Minnesota into Michigan and Wisconsin, the recovery plan requirements were being met rapidly. And as wolves made a comeback throughout the Great Lakes states, the federal government largely ignored the Northeast as potential wolf habitat. It was clear that without a strong advocate, the return of wolves to the sprawling Great North Woods of New York and New England was unlikely.

The Northeast

Defenders of Wildlife has been working on carnivore conservation issues since its founding in 1947. Although the organization works on a variety of wildlife and habitat conservation programs, wolf restoration is its signature issue. For the past 20 years, Defenders has assisted gray wolf reintroduction in the northern Rockies and the Southwest and red wolf reintroduction in the Southeast. Although Defenders biologists were pleased with the ongoing recovery of wolves in the Great Lakes region, for several reasons they felt that restoring the eastern wolf should not stop there. First, a single population is always at risk of being destroyed by disease or other natural disaster, and having multiple populations is a safeguard. More importantly, the Great North Woods of the Northeast contains millions of acres of forested habitat lacking its native top predator. Committed to restoring complete ecosystems and restoring the wolf to as much of its former range as possible, Defenders began its Northeast wolf recovery project in 1994.

At that time there were only a handful of articles about the prospect of restoring wolves to this region (Clark 1971; Engelhart and Hazard 1975; Henshaw 1982). The first step in any species recovery program is to determine the biological and social feasibility of restoration. In the Northeast, there were many questions to be answered. Did the region still contain suitable habitat for wolves? Was the best habitat in Maine, Vermont, New Hampshire, or New York? Could wolves migrate into the northeastern United States from Canada on their own, or would translocation of wolves from Canada be necessary? Would humans want wolves, or at least tolerate their presence?

Recently, several studies have confirmed what wolf advocates have long suspected: Despite wolf populations existing in nearby Canada, wolves are unlikely to disperse into the United States on their own (Harrison and Chapin 1997; Wydeven et al. 1998). Numerous barriers between the northeastern United States and existing wolf populations in Canada hinder possible migration. Expansive urban and agricultural lands separate wolves in Algonquin Park and elsewhere in Canada from available habitat in the northeastern United States. And because wolves are not protected in southeastern Canada, dispersing animals are susceptible to human-caused mortality. Finally, should wolves make it to the border, the St. Lawrence Seaway is yet another barrier to entering the United States. Although the seaway freezes in winter, shipping interests break the ice to allow passage of ships, further reducing the potential for wolf dispersal. Two wolves have been shot in Maine in recent years, but evidence is inconclusive that they had dispersed from Canada.

Because prey density generally was accepted as suitable and the chance of natural wolf recovery was slightly greater in New England than in New York, Defenders opted to start by investigating the feasibility of restoring wolves to the Adirondack Park. A Defenders study (Hosack 1996) indicated that the Adirondacks could hold a small wolf population, suggesting that a more thorough investigation was warranted.

Human Attitudes

Of equal importance to determining the biological feasibility of a proposed restoration project is conducting education and outreach and assessing public opinion on the issue. In 1996 Defenders hired Responsive Management (RM), a Virginia-based polling firm, to conduct a survey of attitudes towards wolves and wolf restoration among people in New England, throughout New York State, and within the Adirondacks. To understand the full range of public attitudes and concerns and develop the survey, RM conducted four focus groups: one in Albany, two in Indian Lake (one made up solely of hunters, a key stakeholder group), and one in Saranac Lake. The information gleaned from the focus groups helped RM create a survey on environmental and wildlife issues, particularly wolf restoration. In addition, RM and Defenders consulted with academic experts Mark Sagoff of the University of Maryland and Willet Kempton of the University of Delaware, who provided feedback and recommendations for survey design. The resulting survey comprised more than 80 questions (Duda 1995). Telephone interviews were conducted with more than 1,200 randomly selected people: 501 in the Adirondacks, 506 throughout New York, and 200 in New England.

The survey was completed in the fall of 1996. Eighty-five percent of New Englanders, 80% of New Yorkers, and 76% of the Adirondackers polled supported wolf restoration into the Adirondacks. Although poll after poll throughout the United States has demonstrated that the majority of the American public supports wolf and other species recovery efforts, the overwhelming results of the RM Adirondack wolf poll were surprising. Today the same survey might yield different results. The RM poll was conducted before the wolf restoration issue came to the forefront of the political scene, and now that groups opposing wolf recovery have become more vocal, public opinion may have shifted in that direction.

Wolf Controversies

At the same time, in November 1996, Defenders sponsored a large wolf conference in Albany, New York. The largest conference of its kind, the Albany meeting attracted more than 500 participants: researchers, educators, wolf advocates, and the general public. The 2-day event attracted extensive media coverage: Newspapers throughout the East Coast and the nation covered the event, as did regional television stations. For better or worse, the wolf restoration issue was launched into the public spotlight. There is a long tradition of conflict on environmental issues within the Blue Line of the Adirondack Park, so it was no surprise that public reaction to the idea of wolf recovery was strong among supporters and opponents alike. With the positive press of the wolf conference and the well-publicized poll indicating that New Yorkers wanted wolves back in the Adirondacks, some local residents who were opposed to the idea became more vocal. Almost instantly the issue became polarized, and resistance to the idea was as ardent as any antiwolf sentiment seen in the western United States over the Yellowstone wolf reintroduction. Suddenly, newspapers all over New York and New England were flooded with editorials and letters to the editor, many supporting wolf restoration but the vast majority voicing opposition. Wolf proponents clearly needed a voice in the Adirondack region, so Defenders hired a field representative to conduct education and outreach exclusively in the north country. A Defenders education coordinator in the Utica region continued to make appearances throughout central New York. Additionally, Defenders staff began meeting with as many stakeholders as possible. At times the heated discourse turned ugly. For example, at the aforementioned Plattsburgh debate, the first author was harassed and eventually escorted from the auditorium by a security guard. The New York Department of Environmental Conservation (DEC) biologist who had attended the debate exited the auditorium to find the tires on his state vehicle slashed. A later meeting was canceled because of a bomb threat.

Feasibility Study

A key message that was getting lost in the frenzied debate was that Defenders was merely advocating the study of wolf restoration, not the immediate release of wolves into the park. It was imperative that a biological feasibility study be conducted to assess the park's ability to hold wolves. However, detractors of the idea had accused Defenders of biasing the Responsive Management study results by funding it, despite RM's outstanding reputation for conducting natural resource and consumptive use polls. The scientific feasibility study had to be credible to all parties, regardless of its outcome. Defenders raised more than $100,000 to fund a study but wanted a neutral third party to oversee the process. Several Adirondack stakeholder groups suggested that a Citizen's Advisory Committee (CAC) be formed to fill this role, and Defenders approached Paul Smith's College of the Adirondacks, a respected natural resource and hotel management school located in the heart of the Adirondacks, about the idea. Jim Gould, vice president for development at Paul Smith's, was interested in the wolf issue and agreed to act as chair for the advisory committee. Jim was well suited for the job, having a knowledge of the issue and the Adirondacks but no real stake in its outcome. In addition, Jim's sense of humor proved invaluable for getting the CAC through difficult discussions. Defenders worked with Paul Smith's College and the DEC to identify appropriate people to serve on the CAC. It was important to build a balanced committee that paralleled the broad spectrum of attitudes toward wolf restoration in the Adirondacks, including proponents, opponents, and those who were undecided. It was also imperative that participants be willing to discuss the issue objectively, particularly in a group setting wrought with different opinions. Finally, the committee had to reflect the myriad stakeholder groups in the Adirondack Park, including hunting, trapping, timber, farming, environmental, recreational, and tourism interests and property owners. Jim Gould and the Defenders staff called numerous groups to request their participation. This networking system generated a list of potential participants. People who were recommended by numerous sources were the most logical to serve on the committee. Ultimately, more than 20 people representing that many different stakeholder groups were invited to serve on the committee. Several declined, including one who felt participation would imply support of wolf restoration and others who could not commit the necessary time. Fortunately, however, most of the invited people accepted.

Process

Committee members were charged with developing a list of issues important to their constituencies for examination in the feasibility study. Over time, they would evaluate proposals from prospective contractors, interview the contractors, select one to conduct the study, outline the terms of the contract, and monitor and provide input to the study. Defenders of Wildlife took a backseat role and was not a voting member of the committee. Although Defenders staff attended every meeting and participated in group discussions, because they did not hold a voting position they did not influence committee decisions.

Because of the volatile nature of the issue and the long tradition of disagreement among Adirondackers on environmental issues, everyone involved thought the process would benefit from professional facilitation during the first CAC meeting in June 1997. Dr. Mark Lapping, provost at the University of Southern Maine, volunteered his professional experience to facilitate the meeting.

During that first meeting, participants quickly aligned themselves with others who expressed similar views. However, despite a tangible level of discomfort, participants worked together successfully to create a comprehensive list of concerns about potential wolf restoration held by Adirondack citizens and stakeholder groups. Biological concerns centered around the ability of the park's deer herd to support wolves. Human health and safety fears focused on the wolf as a potential disease vector and the likelihood of wolf attacks on humans and pets. Economic issues included the costs and funding sources of wolf restoration, the possible impact of wolf depredation on the state's farming and hunting industries, and the potential for wolves to boost regional tourism. The need for land use restrictions and the ability of residents to control problem-causing wolves were the primary regulatory questions.

This list of concerns was developed into a request for proposals, which was distributed widely to universities and other researchers capable of conducting a feasibility study. Jim Gould received three strong proposals from four institutions and distributed them to the CAC members. The CAC decided to interview the prospective contractors to make their selection, so at their fourth meeting, in February 1998, prospective researchers came to Saranac Lake to present their proposals. The committee members were well versed in the wolf issue at this point and challenged the visiting academics with an array of questions about how the work would be conducted, their personal views on the wolf issue, and their proposed budgets. After presentations and a question-and-answer period, the CAC met again in private to discuss the merits of the contractors. There was much debate, but ultimately the committee unanimously agreed on a course of action. They chose the Conservation Biology Institute (CBI) in Corvallis, Oregon, to conduct the biological component, with Paul Paquet and Jim Strittholt as the principal investigators. Cornell University's Human Dimension Research Unit, led by Tom Brown and Jody Enck, was selected to conduct another attitude survey. CBI would have 1 year and Cornell 18 months (so they could base part of their survey on CBI's results) to complete their work.

Results

As of this writing, CBI and Cornell have completed their studies. In the CBI report (Paquet et al. 1999), the researchers determined that although the Adirondacks can indeed hold a small number of wolves, several significant obstacles to wolf recovery must be addressed before a recovery program is initiated. The researchers found two factors particularly problematic. First, few dispersal corridors exist between the Adirondacks and other wolf habitat, which would impede gene flow and long-term survivability of wolves unless active management was undertaken. Second, the taxonomy of the northeastern canid is being questioned. Researchers in Canada, studying wild canids in southeastern Canada and the northeastern United States, have proposed that the wolf from this region, extirpated from New York and New England but still present in parts of southeastern Canada, was not the gray wolf *(Canis lupus),* as previously thought, but rather the red wolf *(Canis rufus).* This research is preliminary and ongoing, and CBI's study concluded that more genetic research is needed on northeastern canids before any reintroduction program should be considered. The Cornell report concluded that at this point in time, the majority of people statewide support wolf recovery and believe that Adirondackers should have the most say in any decisions (Enck and Brown 2000).

Discussion and Future

The CAC process has been praised by state and federal agencies and the media as a model for community participation in wildlife restoration efforts. Having participated in the process and observed the committee function, Defenders is confident that the group dynamic ensured that qualified researchers conducted the work. Because the public plays such an important role in modern conservation, the CAC process proved to be an important mechanism for including public deliberation and participation in decision making. The committee process has been invaluable not only in developing a framework for how varied groups can discuss contentious issues, but for Defenders to better understand northeastern stakeholder concerns and issues about wolf recovery and more effectively merge advocacy with public involvement.

Yet the process was not without difficulties. Every detail and decision had to be discussed by more than a dozen people, who needed to take issues back to their constituencies for feedback before decisions were made. This made the process lengthy and sometimes arduous. Participants worked hard and volunteered their own time and finances to attend meetings. The committee also received some criticism for deciding not to open meetings to the public or the media. However, their decision was important: They took seriously their responsibility to be impartial and felt that media involvement would encourage grandstanding and compromise objectivity. Finally, the lack of trust between members, especially in the early stages of the process, put people on edge and tempers often flared. Despite these difficulties, the process enabled people to treat the issue in a civil, thoughtful, and deliberative manner not often seen with controversial topics. CAC chair Jim Gould often joked, "We are working together towards a common goal, but we're not sitting around holding hands and singing 'We Are the World.'"

A key to the committee's success was adding an element to the process beyond discussion of the wolf issue. Jim Gould initiated a policy of having a meal before every meeting to give participants time to relax and get to know each other. As Jim quipped, "Hypoglycemia and controversy just don't mix!" More importantly, however, the shared meals enabled participants to sit and talk about their families and other aspects of their lives besides the controversial topic at hand. It became clear over the course of the two and a half years the committee met that the dinner component did more to build bridges and resolve disagreements than one could imagine. In fact, not long ago, when a meeting fell on the birthday of a CAC member, the entire group celebrated with cake and song. Group members often met in the hotel's lounge after the meetings. The topic of wolves usually was not discussed during these social times because members knew their friendships and alliances were fragile. Ultimately, the group members developed a sense of respect for one another that transcended the issue. Many participants felt that this process has paved the way for a better understanding among themselves and their respective organizations so that future discussions on other controversial issues can be less acrimonious.

Although the CAC has served as a model for how divergent interests can work together on controversial programs, it has not stopped the emotional debate surrounding the issue. Two counties within the Adirondacks have passed local laws prohibiting wolf reintroduction despite significant public opposition to passing such laws before facts surrounding the issue were known. Although these laws would be superseded by state law should the DEC decide to move forward with wolf restoration, they are clear in their message: Only those who reside should decide. Outsiders, including Washington, D.C.–based environmental groups and the USFWS, are not welcomed by many local residents. Op-eds and letters to editor continue to appear in local papers, although they have slowed while the feasibility study is under way.

Interest in northeastern wolf restoration by opponents, advocates, wildlife agencies, academics, and the general public continues. The USFWS has announced its intent to make the northeastern wolf a Distinct Population Segment and create a new recovery plan that addresses the Northeast separately from the Great Lakes.

Although New England is promising from a biological standpoint, wolf recovery there faces stiff opposition. The Sportsman's Alliance of Maine has attempted to introduce legislation prohibiting wolf reintroduction to the state. The New Hampshire legislature banned the reintroduction of wolves in 1999. On the other hand, more than 23 groups promoting wolf restoration have formed the Coalition for Recovery of the Eastern Wolf (CREW) to counter the antiwolf movement. Finally, several scientific articles have been published in the last 3 years, providing a more objective perspective (Harrison and Chapin 1997; Mladenoff and Sickley 1998; Wydeven et al. 1998; Paquet et al. 1999).

Defenders continues to do its part to educate people about wolf restoration and organize forums for discussing the topic, for without sound knowledge of the species, its ecological importance, or the human impacts of its restoration, communities cannot make informed decisions. The benefits of wolf restoration in the Northeast abound. Wolves could help keep burgeoning ungulate populations in check, deliver tourist dollars to rural communities by providing inspiration to countless wildlife seekers, and fill a genuine wolf conservation need. However, without public participation and support, wolf restoration will remain a mere debate. Defenders will strive to continue and improve on the example set by the Adirondack CAC with the goal of making wolf restoration a reality to the benefit of all.

Chapter Seven

Overcoming Cultural Barriers to Wolf Reintroduction

Rodger Schlickeisen

It is sad but true that an important part of America's wildlife heritage, especially large mammalian predators such as wolves, continues to exist only or
primarily because of human tolerance. For better or worse, humans are playing the godlike role of determining which species will survive and where.
Relentless human population growth and development and an increasingly
technology-driven society are altering the natural world at an increasing pace,
causing the loss of natural habitat, destruction of natural ecological processes,
and accelerating loss of species. How much of wild nature will remain to
enrich the lives of our descendants depends almost entirely on the extent to
which we can change our attitude toward untamed landscapes and wild
species and adopt a more biocentric philosophy similar to that first champi-
oned by Aldo Leopold five decades ago (Leopold 1949).

Whether or how society will develop a more holistic and environmentally
friendly attitude is not yet clear. But one of the best indicators is our evolving
attitude toward wolves, for historically they have occupied a special cultural
niche in American society as the leading symbol of an evil wild nature, a
demon to be conquered and extirpated as quickly as possible by any means
available. This powerfully negative symbolism came to the continent with the
Europeans who brought with them hundreds of years of accumulated prejudice against the wolf, widely expressed in mythology and folk tales from
Aesop's fables to "The Three Little Pigs" and "Little Red Riding Hood." This
led to the single-minded destruction of wolves and has made reintroducing
wolves to areas from which they were deliberately extirpated one of the most
challenging of all conservation undertakings. If American society, especially
the population living near reintroduction-designated areas, can accept the wolf
as a neighbor, it will be a very positive sign of our capacity to elevate our view
of wild species and adopt a more ecologically healthy attitude toward the natural world.

As president of Defenders of Wildlife, this country's most consistent and persistent advocate of wolf recovery, I have been involved to varying degrees with every reintroduction effort undertaken or proposed. Based on this experience, I hold a somewhat hopeful view of the prospects for both wolf recovery and the overall elevation of society's attitude toward wild nature. However, wolf reintroduction can be successful only if proponents heed the lessons that experience teaches about overcoming cultural barriers erected over thousands of years. Following is my view of these lessons, based on Defenders' experience.

Biologically and Economically Successful Reintroductions

As many of the authors in this volume have observed, wolf reintroduction can occur only if two basic conditions are satisfied: The undertaking is scientifically and financially feasible, and the public wants it to happen.

Wildlife experts' experience with four major wolf reintroductions, especially the two in the northern Rockies, shows that they know how to reintroduce wolves, particularly wild wolves, efficiently and effectively. One of the major marks of success in Yellowstone and central Idaho has been the ease with which the translocated wild wolves have adapted to their new habitat, how quickly they have reproduced and formed stable packs, and how readily they have stayed within their planned relocation areas. They have not harassed the local human population, have tolerated exceedingly well the multiple uses (especially timbering, recreation, and agriculture) evident in the recovery areas, and have preyed on nearby livestock much less than anticipated by the Environmental Impact Statement (EIS) prepared before the Yellowstone reintroduction. The EIS also predicted that the wolves' presence would generate millions of dollars of additional tourism spending for the local economy each year; although no studies have been undertaken to establish the amount, the opinion seems overwhelming among Yellowstone Park officials and area tourism-related businesses that it has indeed happened.

The Yellowstone and Idaho reintroductions are way ahead of schedule and well below budget. Although five releases originally were planned in each of the two reintroduction areas, only two were needed because of the rapid increase in wolf numbers. As long as the prey base is adequate and human interference does not exceed reasonable limits, translocated wild wolves can thrive in appropriate recovery areas. The experience with reintroducing captive-bred wolves is less clear, although the indications are that these can be quite successful, albeit over a longer period of time and with more demanding requirements for reintroduction sites. The one reintroduction that did not work, at Cades Cove in the Great Smokies, appears to be one in which the site would not have been selected had officials known what they know now.

Favorable Public Attitudes

If the first prerequisite for reintroduction success is easily satisfied, the second, public support, is more problematic. There is good news and bad. First, the good: The country's attitude toward wolves is very positive.

Although wolves originally were demonized by European settlers and their descendants and were purposefully extirpated from nearly the entire lower 48 states, polls taken since the wolf was first listed as endangered in the 1970s have shown that saving and reintroducing the wolf have become increasingly popular. Overwhelming majorities of the general population have supported wolf reintroduction since at least the early 1990s. Even in the states where wolf reintroduction has been proposed or undertaken, a consistent majority has been supportive. The charisma of the wolf (probably enhanced by the negative mythology, ironically) helps attract media and public support for rescuing the animal. When the first wolves were reintroduced into Yellowstone and central Idaho in January 1995, the media across the country (and indeed much of the world) reported the event more as a celebration than as news.

Defenders' poll results and our extensive experience with direct mail, educational outreach, and other interactions with the public suggest that people support the wolf's cause for a number of reasons. In their approximate order of importance, these reasons are to protect part of "God's creation"; to save something wild, free, and independent in an increasingly tame world; to rescue a creature that has its own intrinsic right to exist; to restore natural ecosystems so they have the creatures nature intended; to benefit future generations; to restore the evolutionary benefits of predation for other animals; and to "help the underdog" and make up for centuries of unjustified persecution. In our experience, the public does not support saving wolves or other endangered species based on any significant belief that doing so will provide themselves or others with more direct utilitarian benefits. This general public support for the wolf is a precondition to winning approval for wolf reintroductions.

Saving endangered species is by law and practice primarily a responsibility of the federal government. Reintroductions of these species take place under the direction of the federal Endangered Species Act (ESA) of 1973. Although there are 45 state endangered species acts, they are weak and ineffective compared with the federal law. But the federal act is funded and implemented only to the extent that Congress and the Executive Branch have the political will to do so. This is especially true of the most controversial aspects of ESA implementation such as reintroducing wild mammalian carnivores such as the wolf. National public support for saving endangered species, particularly for saving the wolf, is what puts force behind the issue and elevates it politically so that it does not depend solely on favorable administrative interpretation, implementation, and enforcement of the ESA. It is what gives reintroduction pro-

posals the momentum necessary to succeed despite the fact that the enthusi-
asm for wolves of the general public is not uniformly distributed across the
country. This brings us to the bad news.

Initially Negative Local Sentiment

Not surprisingly, the most significant antiwolf sentiment (although still generally a minority of the population) is found in the rural areas where reintroduction is proposed. This sentiment stems in part from continued belief in wolf
mythology. More important, in terms of their impact on wolf reintroduction,
are strongly held misperceptions of what the wolf's return would mean to the
established way of life. Attitudes toward the wolf are thus linked to local attitudes involving a number of important other subjects.

Extensive broadcasting of wolf wildlife documentaries, large-scale promo-
tion of wolf recovery by conservation advocates, and generally favorable coverage by the news media have markedly improved the general public's understanding of real wolf behavior and its impact on the natural ecology and the
human-altered landscape. But it is a mistake to underestimate the influence of
the antiwolf sentiment that some people retain. This is especially evident in
rural areas, where it is often informally instilled as part of the culture since
childhood. And nothing draws it forth faster than a proposal to reintroduce
wolves to these areas.

The most persistent myth is that wolves pose a serious threat to human
safety (the "Red Riding Hood" syndrome). Strong evidence of this fear persists despite the fact that there is no known instance of a wolf in the wild
attacking and killing or seriously injuring a human in this country. Yet this
fear is expressed everywhere that wolf reintroductions have been proposed.

A second belief is that wolves will prey on livestock and, to a lesser degree,
kill dogs and other domestic animals. Because this belief has some basis in reality, the problem is one of exaggeration, which can often include lurid tales of
wolf packs purposefully and without provocation or other specific stimuli
going out of their way to needlessly—indeed, "ruthlessly"—slaughter large
numbers of cows or sheep and kill family pets right on the front porch. The
fear that this exaggeration generates is most noticeable in areas where cattle
and sheep ranching is common and where such predation is therefore perceived as a serious economic threat.

A third belief is that wolves will prey so extensively on wild ungulate game
animals that those who hunt those animals or act as commercial guides to
tourist hunters will see their supply of targets reduced. Again, the problem is
one primarily of exaggeration.

Fourth is the fear that reintroducing the wolf as an endangered species will
bring with it government-imposed restrictions that seriously impinge on his-

toric access to public lands, private property rights, and personal freedoms. Some variation of this fear is found everywhere that reintroduction is considered, although its specific nature varies with the history of the area and the prevailing sentiment toward government. In areas where ranching is significant as a way of life, an influential fear is that the wolf's presence will severely restrict where and when one can put livestock or even go alone and what one can do to protect livestock from wolf predation. Some residents in the vicinity of the Olympic National Park and Adirondack Park, and even near Yellowstone National Park, harbor serious resentment against the government for appropriating the land and establishing and perhaps expanding it as a park. In that context, a government initiative to restore the wolf can be seen as "more of the same" and easily becomes the most serious obstacle to overcome. Like other fears, although it isn't unfounded, it is exaggerated.

Finally, complementing these beliefs and fears is a general perception that those who support the return of the wolf, such as environmental advocates, are dismissing the local population's way of life and values as unimportant or wrong and are even working to end that way of life. Of course, in some areas this is not an unreasonable perception. Some elements of the environmental community are aggressively outspoken in proposing to end both ranching and timbering because of the harm they can cause to the natural environment. Animal welfare and animal rights groups sometimes actively support wolf reintroduction, and they can be vocal in criticizing ranchers, albeit for different reasons. Local residents worried about the continuance of their way of life are unlikely to see any meaningful distinction between prowolf and antiranching (or antitimbering) activism.

Political Opposition

Although the ESA is a national law, the conservation actions it mandates occur one at a time in individual states or regions. Meanwhile, Congress has a tradition of giving great deference to the wishes of each state's congressional delegation on matters that occur within its boundaries. Whether or how much members of the delegation try to draw on that source of power depends heavily on their constituents' wishes and their influence with the delegation.

Because of the political dynamics at work, local residents near proposed wolf reintroduction sites typically have significant influence with their elected officials. One reason is that, as the old saying has it, "the squeaky wheel gets the oil," and local wolf opponents tend to squeak much louder than local wolf advocates. Their perception of what the wolf's return would mean, and the fears that this generates, seems to make them more fiercely committed to their position than are wolf supporters. Just as important, perhaps, is the fact that the wolf opposition most closely reflects the cultural traditions and values of the area and thus fits most comfortably with the historic political positions on wolf and other natural resource issues taken by elected officials representing the area. It is the ideas of environmental advocates that are new and must struggle to exert influence with elected officials.

The result is that the members of the congressional delegation from states proposed for wolf reintroduction have set up major roadblocks by pressuring the federal land agencies involved or thwarting reintroduction progress with provisions they attach to federal legislation (especially appropriations bills that fund the ESA). This influence is easiest to exert in cases where the specific reintroduction is not mandated by the ESA but rather is proposed as a voluntary action to be undertaken by the state with federal agency assistance (the proposed reintroduction of wolves into the Adirondack Park is a current example). Governors and other state officials similarly have influence because of their political connections and leverage with federal officials and because they determine the attitude of state wildlife agencies, which always play a role whether the reintroduction is ESA mandated or voluntary. This influence varies between states and reintroductions, of course, depending in part on the significance of any nonfederal lands proposed as wolf habitat.

Illegal Killing Threatens Reintroduced Wolves

Even where the reintroduction isn't blocked, the battle isn't necessarily over when the wolves leave their crates or pens. As noted earlier, one lesson learned from past reintroductions is that wolves have a great tolerance for the kind and level of multiple-use activities that take place in and around the reintroduction areas. Notwithstanding the fear that the wolf presence will necessitate extensive new land use restrictions, the new restrictions have been minimal. Most often they amount to little more than limiting access to the vicinity of wolf dens in the spring when the pups are born.

At the same time, however, as the earlier history demonstrates, wolves are extremely susceptible to human persecution and are among the easiest large mammals to find and extirpate. Particularly important in this regard is the fact that wolves live and hunt predominantly in family groups in permanent territories: Find one wolf and you probably will find the full pack. For those determined to kill them, the task is easy. And the motivation to act is there if the combination of fear and resentment is strong enough and there is cultural support for taking matters into your own hands (as exists in the "shoot, shovel, and shut up" commandment popular with the most aggressive wolf opponents in the northern Rockies and the Southwest). For this reason, it is a matter of consensus among wildlife professionals, wolf advocates, and others with direct reintroduction experience that the biggest threat to wolf viability is illegal killing by disgruntled local residents.

Addressing Local Concerns

Myths

There does not appear to be any way to eliminate local opposition or the threat of illegal killing once reintroduction takes place. However, it is possible to improve local acceptance and reduce the threat of illegal killing. Indeed, if this is not done, wolf reintroduction will not go forward or, if it does, will not be successful.

Fortunately, one of the biggest single problems, mistaken perceptions about the threats that wolves pose, can be addressed directly. The perception that wolves pose more than the most minimal threat to humans or that wolf predation will decimate the local livestock herds or wild game population can be corrected by bringing credible, unbiased experts directly into the local communities. There they can give lectures and slide shows, appear in panel discussions open to the public, participate in radio and television interview shows, and conduct editorial board visits with influential area newspapers.

Two outreach formats have proven particularly effective. One that was used successfully on the Olympic Peninsula is to conduct a full-day workshop filled with panels of experts on all aspects of wolf reintroduction and to invite representatives from all segments of the community, including opposition groups, to attend, listen, and ask questions. The local media should be invited as well.

A second effective outreach format is to sponsor a tour of wolf experts (such as Mission: Wolf and Wild Sentry) with their real wolves through the region that includes the proposed reintroduction area. There is nothing quite as effective at debunking mythology's "bloody tooth" image of wild wolves as seeing real wolves eye to eye. (At least this is true with a large percentage of the population; the most aggressive opponents can remain as solidly antiwolf as before and can be further incensed by the tour itself.) The format works equally with schoolchildren and adults and always can be counted on to produce extensive local print and television coverage that further spreads the educational message. Sponsoring educational booths at popular locations and providing wolf curricula to teachers also are effective, especially when they complement live wolf tours.

In the educational outreach effort, it is extremely important that spokespeople be absolutely honest and objective in their presentations, resisting the temptation to counter the exaggerated claims of wolf opponents with exaggerated counterclaims. Such an effort to "balance" the two presentations and hope that the target audience settles on a view somewhere in between will backfire. If the audience is initially inclined to give wolf opponents the benefit of the doubt and if wolf advocates are revealed to be spreading untruths, that may be the end of the debate as far as the local community is concerned. The truth about wolves, even if it won't qualify them for sainthood, is sufficiently unthreatening to remove or minimize the opposition of most of the local citizenry. And wolf advocates must be seen as being credible, or political opposition will remain insurmountable.

Of course, one result of being honest is that the public will come to recognize that although wolves represent a minimal threat to humans and pets and will not decimate healthy populations of wild ungulates (there is always a reintroduction provision providing for wolf control if the wolves push down ungulate populations to unacceptably low levels), they do represent a measurable and potentially unacceptable level of threat to livestock.

Livestock Depredation

An important part of conducting outreach to local communities is to explain that because wolves have been genetically programmed by thousands of years of experience to prey on wild game, that is what they will generally do. But it is necessary to acknowledge that on rare occasions, wolves will kill livestock. Furthermore, because it is a smart animal, once a wolf preys on cattle or sheep and thus learns how easy that is compared to killing wild game, it can quickly adopt it as a habit, which it imparts easily to the rest of the pack.

Understandably, most ranching communities regard wolf predation as unacceptable because it poses a threat to their economic well-being. They will reduce their aggressive opposition to reintroduction only in response to a credible promise that reintroduced wolves will be managed in a manner that ensures that livestock predation will not get out of hand and that they will not suffer economically.

Although it isn't possible to remove all livestock predation or provide complete assurance that ranchers won't incur some financial loss, it is possible to minimize both. First, to keep livestock predation from becoming a widespread habit among the wolves, which can incite local ranchers to kill them illegally and defeat the reintroduction effort, wolves that prey on livestock must be quickly and effectively controlled. They may be relocated away from livestock once or even twice, but if they continue to kill livestock they must be removed from the wild or killed. (A current Defenders initiative involves searching for effective nonlethal, preventive wolf control techniques.) This is probably the single most difficult lesson for wolf advocates, and especially for animal welfare and animal rights groups, to learn: To win tolerance for the majority of wolves that do not prey on livestock, it is necessary to support control of the few wolves that do.

Quick, effective wolf control can minimize livestock losses. Particularly when viewed from the perspective of total annual losses of livestock, predation by wolves is minuscule. (Although wolf recovery in northwestern Montana has been under way for more than a decade and wolves were returned to Yellowstone and nearby central Idaho beginning in January 1995, total annual livestock losses to wolves in the region are counted in the dozens, while losses to coyotes number in the tens of thousands and losses to domestic dogs easily exceed a thousand.) But this small overall loss is significant for two reasons: First, it is important to the handful of ranchers who actually experience it, and second, it is important to the public, local and national, who view it as unfair that one small group of people should have to bear alone whatever financial burden there is to returning wolves to public lands. Together, these two factors can make wolf predation the most powerful argument against reintroducing wolves into ranching areas.

The solution is simple and straightforward: Compensate the ranchers for any verified livestock predation caused by wolves. One of the reasons for the success of wolf recovery in Minnesota is the fact that the state government compensates ranchers for predation by wolves. And those who know the specifics of the successful wolf reintroduction into Yellowstone National Park and central Idaho credit the compensation program run by Defenders of Wildlife with making the reintroduction possible and minimizing subsequent illegal wolf killing. Defenders' extension of the compensation program to the Southwest is credited with facilitating wolf reintroduction there in January 1998.

In addition, whatever compensation program is initiated must be viewed as serious, fair, and dependable by the ranching community. Defenders has attempted to accomplish that by having the determination of each claimed wolf kill be undertaken not by itself but rather by professionals employed by the U.S. Department of Agriculture's Wildlife Services program, which is viewed very favorably by ranchers; paying ranchers for verified losses at the projected market price the livestock would have commanded had it been sold in the fall, and often paying half or all of market value even when losses are suspected but cannot be verified; mailing checks within a few days of when a claim is verified; and providing a special compensation trust always maintained at $100,000 or more to pay for wolf predation claims. Finally, it isn't enough just to have a compensation program; it must be advertised. That is the only way it can positively influence local and national public opinion.

Property Rights and Personal Freedom

Remaining unaddressed thus far are perceived threats to property rights and personal freedom. To improve local acceptance of wolf reintroduction, proponents must act in a manner that respects the local culture and design the reintroduction to accommodate the local perspective as much as possible.

As noted earlier, experience has demonstrated that reintroduced wild wolves can easily tolerate most if not all existing land use activities and still thrive. Aside from barring access to areas around den sites in the spring, few other land use restrictions are necessary.

Experience with reintroduced captive-bred wolves is more limited and precludes generalization. Although the reintroduction in the Southeast with captive-bred red wolves appears to have been successful with no land use restrictions, the reintroduction of 11 Mexican wolves into Arizona in early 1998 resulted in several incidents in which campers in the immediate reintroduction area encountered wolves that did not demonstrate the aversion to humans typical of wild wolves. The obvious assumption is that their prior exposure to humans throughout their lifetimes left them unafraid and unwary. One wolf was shot by a camper who claims to have been frightened for his family's safety, although there is no evidence the wolf was threatening the family (O'Driscoll 1998). Other wolves have been shot illegally by unknown assailants, and one can only speculate that they were easy victims because of their unwariness. (Sizable financial rewards have been offered for information leading to the arrest and conviction of wolf killers, and as of March 1999 there were 2 wolves in the wild and reintroduction of 11 more planned.) Because future reintroductions are likely to involve translocating wild wolves, the prospects are good for accomplishing them without land use restrictions, which the local community might regard as punitive.

Perhaps the most serious of the perceived threats to established values is the expectation by local ranchers that they will not be permitted to protect their livestock from wolf predation. Of course, the tradition of ranchers being allowed (indeed, encouraged) to kill animals that prey on sheep and cattle is very firmly established in local cultures. In fact, ranchers tend to regard shooting wolves, coyotes, or other predators attacking their livestock as a matter of simple common sense. This practice has existed since the land was first settled, in many cases by ranchers' own grandparents or great-grandparents.

It is very difficult for some environmentalists and most animal welfare and animal rights advocates to accept, but experience with the two reintroductions in the northern Rockies shows that allowing ranchers to shoot predators caught in the act of preying on livestock on the rancher's own property is advisable to reduce both local opposition to reintroduction and illegal killing once reintroduction occurs. It also results in few wolf deaths, probably fewer than would occur in its absence. This is because wolves that prey on livestock eventually would have to be killed by wildlife authorities even if the rancher did not do so. Also, if the rancher shoots the wolf in the act of predation, it eliminates the wolf quicker than wildlife authorities could, thus reducing that wolf's opportunity to teach livestock predation to the rest of its pack, perhaps sparing the whole pack from being killed later.

Section 10(j) of the ESA, the "experimental population" provision, allows reintroduction protocols to be flexibly tailored to accommodate just such needs (Endangered Species Act 16 USC 1539j). On January 13, 2000, the federal Tenth Circuit Court of Appeals confirmed the legality of such use of Section 10(j) (*Wyoming Farm Bureau v. Babbitt* 2000). The Farm Bureau, which wants to stop all forms of reintroduction including those that accommodate local values, and various wolf supporters who want reintroduction only on a basis that rejects those values (presumably in the interest of better protecting the wolves), claimed the provision had been too liberally interpreted in the northern Rockies. The unanimous ruling in the Tenth Circuit, not known to be particularly favorable to environmental laws, should ensure that appropriate reintroduction projects can continue to use 10(j) to accommodate local values without running afoul of the court's interpretation of the law's intent.[1]

Finally, a logical extension of improving acceptance by accommodating local values is to actively involve representatives of the local community in developing the reintroduction protocol. Federal wildlife officials do this as a matter of course with their extensive schedule of hearings, solicitation of comments, and consideration of multiple analyses and plans. But wolf advocates from the nonprofit environmental community can and should similarly reach out to the local community to solicit their views and, where practicable, include them in joint efforts to design workable reintroduction plans. Although it may be impossible to work constructively with people who oppose wolves under any circumstances, wolf supporters must strive to identify local citizens who are willing to engage in reasoned dialogue. Only by making such an effort can we hope to get past dealing with just the handful of hot-button issues that cause polarization and draw on the base of shared environmental values that make living in the area appealing to all. Discovering those shared values can promote a willingness to work together.

Reaching out to the local community helps advocates understand local perspectives, which often can be accommodated through use of Section 10(j) flexibility and other means such as predation compensation. These measures also demonstrate respect for the established way of life, which dissipates some of the community's preconceptions about the attitude and intentions of environmentalists. In its proposal to reintroduce the eastern timber wolf to New York's Adirondack Park, Defenders is taking the outreach much further, sponsoring the formation of an advisory group of local citizens who represent the full range of positions on reintroduction, from solidly positive to equally negative, to consider and guide the analytical phase (see Fascione and Kendrot, Chapter Six, this volume). It is too early to speculate on the ultimate outcome. However, the collaborative undertaking has produced a fairly cohesive working group that has already designed the basic feasibility study, distributed a request for proposals, and selected two consulting groups to undertake separate aspects of the study. This progress bodes well for the future. (Although it does not involve wolves, there is a similar experiment in Idaho using a voluntarily formed citizens' advisory group working on a controversial grizzly bear reintroduction proposal. A reintroduction protocol designed by the citizens' group has been selected by the U.S. Fish and Wildlife Service as the preferred alternative for reintroducing the grizzly, and the secretary of the interior is expected to choose that alternative in 2000.)

Persistence and Strategic Flexibility

Even with the most productive efforts to improve local acceptance of wolf reintroduction, it remains necessary to continue to build, reinforce, and demonstrate support among the national media, the general public, and sympathetic elected and appointed officials. At this stage in the evolution of society's attitude toward big, wild carnivores, the commitment to reintroductions remains fragile. The decision to go forward with any proposed reintroduction is largely a political one, and until wolves have actually been reintroduced—and sometimes even after they have been—the project is susceptible to shifting political events.

It is well to keep in mind that the path of least resistance for politicians is to delay, block, or even terminate planned or ongoing reintroductions to appease one or more groups that happen to have political influence.

Therefore, reintroduction advocates must remain ever vigilant, continue to cultivate proponents and champions to give the cause support, and step forward quickly when the attacks occur. A prime example of just such an attack occurred in August 1995 when Senator Jesse Helms, without any warning, offered an amendment to the Department of the Interior Fiscal 1996 appropriations bill to terminate the red wolf reintroduction program in North Carolina. With less than 2 hours available, environmental activists and prowolf officials within the Executive Branch rallied aggressive opposition from several Senate champions, who were able to defeat Helms by 2 votes, 50 to 48. Although it was distinguished for having been the Senate's first proenvironment vote in the 104th Congress, the narrowness of the victory is indicative of the slim margin reintroduction advocates enjoy in Congress.

Sometimes the political forces work in the opposite direction. Although the gray wolf was first listed as endangered when the ESA was passed in 1973, nearly 15 years passed before agencies of the Department of the Interior were willing to promote reintroduction and longer to fight past the blocking and delaying tactics used by members of the congressional delegations from Idaho, Montana, and Wyoming. Advocates had to use everything from reason to legal action to accomplish the objective. (Even after wolves had been released for 2 straight years, in 1995 and 1996, congressional opponents continued to oppose the reintroduction by defeating additional appropriations. In 1996 it appeared that they would be successful in blocking funding for the third release, planned for the following year. Fortunately, the first two releases had been successful enough that wolf advocates agreed that the project probably could succeed without additional ones and that it would be a wise strategy to withdraw the funding request and avoid a political showdown that could have imperiled the whole program.)

Because wolf reintroductions are so political, the prospects for moving forward vary with the shifting political landscape. The members of the relevant congressional delegation are key, but so too are the subject state's governor and wildlife officials, the president and the secretary of the interior, and the head of the Fish and Wildlife Service and sometimes those of the Forest Service and National Park Service. Just as wolf advocates must be wary of political changes bringing new threats, they also must be alert to political changes that open windows of opportunity. Wolf reintroduction was stymied by the Reagan and Bush administrations, but the Clinton administration brought more sympathetic officials who were encouraged to move forward and did so in the northern Rockies and the Southwest.

Conclusion

The biggest problems confronting wolf reintroduction are not biological or ecological but cultural. Although the general public is supportive, significant portions of the rural community in the designated reintroduction areas are likely to perceive wolves as threats to physical security, economic welfare, and personal freedom. Wolf opponents aggressively exploit these perceptions by generating opposition among elected officials with whom they often share negative attitudes about predators and related issues. To overcome such culturally based opposition, advocates must undertake extensive outreach in the rural community. The prospects for success increase significantly if wolf advocates show respect for the rural residents' cultural perspectives and respond with education to counter negative mythology, with financial assistance to compensate for predation on livestock, and with flexibility to limit regulatory requirements to those that are necessary. Success in these efforts can facilitate reintroduction and reduce illegal wolf killing after reintroduction. Finally, wolf reintroduction is also a political decision, subject to the vicissitudes of the changing political landscape. Wolf advocates must respond with vigilance and persistence and demonstrate their own political power and strategic flexibility to overcome the hurdles.

Endnote

1. On January 13, 2000, the U.S. Court of Appeals Tenth District reversed the lower court ruling, finding that the experimental designation for wolves reintroduced to Yellowstone did not violate the ESA (*Wyoming Farm Bureau v. Babbitt* 2000).

Chapter Eight

In Wolves' Clothing: Restoration and the Challenge to Stewardship

Jan E. Dizard

For much of the past 150 years, roughly the period in which the genre of nature writing developed and during which modern environmentalism took shape, Americans who cared to listen have heard a mounting litany of woe. George Perkins Marsh (1965) set the tone in the mid-nineteenth century when he observed that our ignorance and indifference were upsetting the delicate balance that gives the physical world its integrity. In our wake, Marsh saw a turbulent mix of disruption and ruin. On his world travels and from his farm in central Vermont, he watched thickly forested hillsides denuded, replaced by farms and grazing domestic animals. The trees that once shielded the land were turned into lumber and fuel, the latter filling the air with acrid smoke. The spread of agriculture prevented reforestation, depriving wildlife of vital habitat at the same time that wildlife populations were being heavily exploited for meat, fur, hides, and feathers. The physical world around him was threatening to become Humpty Dumpty, shattered into a jumble of fragments lacking in coherence. Ever since, nature writing has been framed by this narrative of loss. Even when the writing was rhapsodic, as was Muir's, the spectacles and wonders evoked could not help but impress upon the reader the widening circle of disruption and degradation against which Muir's Sierras seemed even more compelling. This sense of loss lent urgency, in Muir's day as in our own, to the desire to protect the few places that had not borne the full brunt of civilization.

To be sure, there was a counterpoint to writing about nature in elegiac mode. John Burroughs, a contemporary of John Muir and the most widely read nature writer around the turn of the twentieth century, wrote compellingly about the marvels close observation of nature revealed, including even the highly modified nature of one's own backyard or nearby vacant lot. Although Burroughs traveled widely, unlike Muir he did not insist on making invidious distinctions between nature in farm country and the pristine high Sierras. Burroughs insisted that nature is everywhere and on all scales irre-

pressibly bringing life forth. Our backyards cannot compare with the specta-
cle of Yosemite Falls, but the delicacy of a moth's wing is every bit as spell-
binding at home as it is in some sublime setting.[1] He was an ardent supporter
of the Adirondack Park, at least in part because it embraced a mix of the sublime and the prosaic rather than elevating one at the expense of the other.

Echoes of Burroughs's inclination to temper loss and degradation with
accounts of nature's resilience continue to the present, even though they are
often embedded in the larger narrative of loss. John Mitchell's (1997) chroni-
cle of the environmental ups and downs of 1 square mile in New England fits
comfortably in this tradition, reminding us that nature is as fully alive under
our noses as anywhere else.[2] Mitchell writes,

> Wilderness and wildlife, history, life itself, for that matter, is some-
> thing that takes place somewhere else, it seems. You must travel to
> witness it, you must get in your car in summer and go off to look
> at things which some "expert," such as the National Park Service,
> tells you is important, or beautiful, or historic. In spite of their
> admitted grandeur, I find such well-documented places somewhat
> boring. What I prefer . . . is that undiscovered country of the
> nearby, the secret world that lurks beyond the night windows and
> at the fringes of cultivated backyards. (Mitchell 1997, 7)

Robert Sullivan (1998) shares this fascination, although he is drawn to the
scene of desolation that adds a deeply ironic note into his encounter with
nature. Sullivan's chosen habitat is a most unlikely place in which to appreciate nature: The Meadowlands, just across the Hudson from New York City, is
one of the most alternately abused and neglected patches of territory in the
New World, right up there with Butte, Montana, and Sudbury, Ontario. Even
a talent as large as Burroughs would have had trouble finding anything inspiring, much less redemptive, in the Meadowlands. Yet Sullivan finds nature at
work in places seeming to have neutralized if not erased the most obvious evidence of our abuse. Like salmon defying currents and gravity to get upstream,
microbes team up with flora and fauna to reclaim what they can. There are
impediments to this reclamation, of course. However generative and resilient
nature might be, we have put some impressive things in nature's way, just as we
have erected dams for which the salmon is no match. The Meadowlands will
never be what it was before we began dumping, filling, and otherwise defiling
the area. But, Sullivan implies, if we can keep from insulting it further, if we
could even give nature a bit of a helping hand, the Meadowlands might yet
teach us something about defilement and the regenerative powers of nature.

These two narratives, one of loss, the other of recovery, have dominated our
thinking about nature and have shaped our contradictory and often embattled
attempts to define an ethically and materially sustainable relationship with the

natural world. In their extreme versions, these two narratives are diametrically opposed. The narrative of loss becomes a wholesale condemnation of modern society and an evocation of a fast-approaching apocalypse (McKibben 1990; Berry 1988). At the other extreme, the narrative of recovery can provide a fig leaf of respectability for the so-called Wise Use movement and others who reject almost all environmental regulation and restraint on our exploitation of nature. The narrative of recovery also underpins the views of those few scholars who remain convinced that nature's bounty is unlimited, capable of absorbing very large increases in human population (Simon 1981).

Fortunately, we do not have to be captive of either narrative—at least not yet. Reckoning with our many losses, we have begun to develop and refine ways to reclaim and restore habitats and reintroduce wildlife species to habitats from which they have been driven. Many promising efforts are under way to boost the recuperative capacities of nature. *New York Times* science writer William K. Stevens (1995) has chronicled the efforts of a group of biologists and environmental activists, marching under the banner of restoration, who are intent on bringing back as much of the original prairie as current land use can accommodate. Two other science writers, Stephen Budiansky (1995) and Gregg Easterbrook (1995) have made even stronger cases for what can be done when the recuperative powers of nature are harnessed to science and careful environmental and resource management technologies.[3]

The idea of restoration is not exactly a new one, although the scope of many recent restorative efforts has given restoration a new cachet (Baldwin et al. 1994). For example, restoring wildlife populations has been a goal of the state and federal agencies setting wildlife policies for many decades. At first, efforts were directed almost entirely at reclaiming habitat and restoring wildlife species prized by those who fished and hunted. Indeed, protecting game species from overexploitation and replenishing the stocks of species that had been overharvested was one of the main motivations for the environmental policies of Theodore Roosevelt and the early promoters of conservation (Reiger 1975). Continuing this tradition, in the 1960s state and federal wildlife agencies began concerted efforts to restore wild turkeys to much of their former range, and in most of these areas populations are now robust; in fact, in some places people are beginning to complain of "nuisance turkeys." Migratory waterfowl numbers were heading precipitously downward in the mid-twentieth century, largely because agricultural practices were destroying the wetlands the birds needed for nesting. Government and private efforts have gone far in slowing the loss of wetlands. As a result, the populations of many waterfowl species are at least stable and some have grown. Indeed, the rebound of one bird, the snow goose, has been so great that they are threatening to displace many other species of geese and ducks.

Modern management practices of these sorts have combined with economic and social changes to produce, often inadvertently, prime habitat for many animals whose presence close to human settlements poses a new range of challenges. New England is experiencing a steady increase in collisions between automobiles and moose as the moose fan out from the sparsely settled north woods into densely settled areas to the south and west.[4] Black bear populations are also increasing and dispersing into suburbs where most residents have known only teddy bears.

Game animals are no longer the only creatures whose numbers and range are being deliberately augmented. Spurred by the Endangered Species Act (1973), nongame species have come closer to their rightful place in public wildlife policy.[5] The bald eagle, the peregrine falcon, and the California condor have been brought back from the brink of extinction. Of course, we can't claim credit for every instance of resurgent wildlife. Despite our determined efforts to purge the landscape of coyotes, they have increased their range and numbers dramatically. They are now abundant throughout New England, apparently filling the void left when we exterminated the larger carnivores with whom the coyote could not compete. Similarly, beaver populations in New England are increasing rapidly with only modest deliberate assistance from humans. With an abundance of trees and few predators, the net effect of our unconscious preferences and land use patterns, beavers are repopulating the New England landscape.

Although this resurgence is encouraging, no one should imagine that the cumulative effects of thousands of years of human appropriation of nature can be undone. Many life forms have gone extinct by virtue of our proliferation, and it is clear that even with our best intentions and concerted efforts, many more species will disappear by our hand. This will continue to fuel the sense of loss and, for some, give urgency to the desire for recovery.[6] Herein lies the danger. If we are too indiscriminate in our efforts, we will exhaust ourselves by trying to do the impossible; if we are too narrowly selective, if we focus our efforts only on a few charismatic species and inspiring habitats, we run the risk of making things far worse than they might otherwise be. Hubris is a source of grief, whether it leads us to push creatures away or to bring them back. This turns as much on public perceptions as it does on natural forces. Right now, public perception clearly favors bringing the wild back, in most if not all its forms. How long this will be so remains an open question.

Humpty Dumpty's Great Fall: The Public Sense of Environmental Loss

One of the truly remarkable features of an otherwise polarized and volatile public has been the consistency of the American public's concern over the environment. For the past 25 years, public opinion polls have recorded strong support for spending tax dollars on the environment. For example, the National Opinion Research Center (NORC) at the University of Chicago has been asking carefully drawn random samples of American adults about their support for environmental spending nearly every year since 1973. In only 2 of 25 years between then and their latest survey, completed in 1998, did support for increased spending on the environment dip as low as 50%. In most of the past 25 years, two-thirds to three-quarters of Americans said that the United States was spending too little "on improving and protecting the environment." This willingness to commit more tax dollars to the environment is all the more impressive given the prevailing antitax, anti–government spending mood of the country. Indeed, public support for environmental spending rose sharply through the Reagan–Bush and Bush–Quayle era.

It also should be noted that support for spending on the environment is evenly spread throughout the population; there is no gender gap, and differences along racial lines are small compared to black–white differences on most other policy issues. Only when we look at age do we find appreciable variations in levels of support for increased spending on the environment. As one might expect, young Americans are far more likely to think we are spending too little on the environment than are older Americans: 74% of those under 30 say we are spending too little, whereas only 47% of those 50 years old and older think too little is being spent.

In 1993, 1994, and 1996, NORC asked a range of more detailed questions about Americans' attitudes toward the environment, and these questions allow us to go well beyond the public's general endorsement of higher spending on the environment. More than half of the respondents in 1993 and 1994 claimed that "I do what is right for the environment, even when it costs more money or takes up more time." Once again, the only variation of note in this response comes with age, but here things are reversed: Surprisingly, the younger respondents are less likely (46%) to "do the right thing" than their older counterparts, 63% of whom say they try to do what's right.

To be sure, all is not sweetness and light. When the questions get more pointed, support for the environment starts heading south. Barely 50% of those polled in 1993–1994 said that they would be willing to pay "much higher prices," a little more than a third said they would be willing to pay "much higher taxes," and less than one-third said that they would accept cuts in their standard of living "in order to protect the environment." It is tempting to interpret this cynically and conclude that because we are not prepared to put our money where our mouth is, there is little reason to pay attention to our pious utterances. Still, however important money is, it is not the only measure of commitment to protecting the environment. For example, when they were asked in 1996 whether they agreed that "natural environments that support scarce or endangered species should be left alone, no matter how great the economic benefits to your community from developing them commercially might be," 60% of the respondents indicated agreement. Unfortunately, we do not have a time series on this question, but it is hard to imagine a representative cross-section of Americans giving assent to this question in, say, 1950. Even more impressive, especially with all the teeth gnashing over government intrusion on the rights of individuals, especially their property rights, that has characterized public discourse for the past 20 years, Americans are surprisingly willing to accept regulations aimed at protecting the environment. For example, 89% endorsed the following statement: "It should be the government's responsibility to impose strict laws to make industry do less damage to the environment." More astonishingly, 73% agreed that "for certain problems, like environmental pollution, international bodies should have the right to enforce solutions." It appears that only a handful of Americans are worried about the United Nations flying unmarked black helicopters over Idaho and Montana.[7]

If this were all there was to the story, matters before us would be simple: Americans, by a large margin, endorse solicitude for the environment, even though their resolve wavers when that solicitude takes something out of their pocket or puts a crimp in their lifestyle. The plot thickens, however, when we consider what people have in mind when they say they embrace nature. The shift in the way Americans view nature goes well beyond wanting more trees and fewer strip malls. Appetite for the wild has been awakened, almost as if large numbers of our fellows had read and taken to heart Thoreau's memorable assertion, "In wildness is the preservation of the World."

Thoreau was not speaking for his neighbors when he praised wildness. They were still busy pushing wildness as far away as possible, preferring clear boundaries, well-kept fields, and providently managed woodlots. By the beginning of the twentieth century, perhaps because the prospect of wildness holding back the march of civilization no longer seemed threatening, steadily growing numbers of Americans sought out remote areas that, though not wild in the fullest sense, stood in sharp contrast to the cities and suburbs from which the sojourners hailed. Hiking and camping, as well as fishing and hunting for recreation as opposed to subsistence, became very popular and once paid vacations became widespread, visits to state and national parks and forests rose sharply (Schmitt 1969).

Thoreau's wild was an abstraction, a metaphor really.[8] Ever since Thoreau, people drawn to the wild have imagined that by immersing themselves in nature, they come face to face with eternal truths, not least of which is to witness how the natural world is intricately stitched together in elaborate webs of symbiotic relationships. The wild has thus come to stand as a symbol of all that is pure and uncontrived, a sharp contrast to the landscapes we have tamed.

In the early stages of this shift in perspective, the desire to preserve at least portions of what remained of the wild got folded in, albeit with some creases, with the emerging ethos of stewardship. The idea was to conduct ourselves in ways that would sustain natural diversity and yield a continuing harvest of resources, both material and aesthetic: Preserve here, prudently use there, and, where possible, promote the idea of multiple use to broaden the ranks of those who have a commitment to the wild.

Along the way, in ways that are still not well understood, Americans began to see nature less as a storehouse and more as something precious that should be protected, treated more like a museum than a storehouse. Stewardship became suspect because it was joined at the hip with the consumptive use of nature. The narrative of loss turned the dominant national celebration of Manifest Destiny and growth on its head; the march of progress was headed for a cliff, and the state and federal agencies managing our natural resources seemed all too willing to run interference for the exploiters. Again, data from NORC's General Social Surveys of 1994 and 1996 are instructive.

A solid majority of Americans in the 1994 and 1996 surveys agreed with the statement, "Almost everything we do in modern life harms the environment." More importantly, Americans appear to have lost confidence in our capacity to solve environmental problems by relying on science and technical know-how. Only 20% of the respondents in both surveys thought that "modern science will solve our environmental problems with little change to our way of life." Moreover, 55% agreed with the statement, "We believe too often in science, and not enough in feelings and faith." And, as if to emphasize the importance of faith, nearly 80% said that "human beings should respect nature because it was created by God." Given this set of beliefs, it should not be surprising to learn that just over half of all respondents reported believing that "nature would be at peace and harmony if only human beings would leave it alone."

An even more nuanced view of Americans' perceptions of nature and the environment can be gleaned from the work of a team of anthropologists (Kempton et al. 1995) who interviewed several hundred people chosen from five distinct groups: members of EarthFirst!, members of the Sierra Club, the "general public," workers in the dry cleaning industry, and workers in sawmills.[9] As one would expect, there are some dramatic differences between these five groups of respondents, but what is even more impressive is the degree to which there is consensus on precisely the matters we have been discussing. For example, almost everyone agreed that "we have a moral duty to leave the earth in as good or better shape than we found it." Sawmill workers were the only group in which fewer than two-thirds agreed that "nature is inherently beautiful. When we see ugliness in the environment, it's caused by humans." (Only one-third of the sawmill workers agreed.)

Ranks closed again on the statement, "Nature may be resilient, but it can only absorb so much damage"; 94% of EarthFirst!ers agreed, as did 85% of sawmill workers. Similar accord was found on the statement, "Nature has complex interdependencies. Any human meddling will cause a chain reaction with unanticipated effects"; 97% of EarthFirst! members agreed, as did nearly two-thirds of sawmill workers (63%). Sawmill workers again split from the pack on the question of extinction. Large majorities, ranging from 78% to 97%, agreed that "preventing species extinction should be our highest environmental priority. Once an animal or plant species becomes extinct, it is gone forever." Only 41% of the sawmill workers agreed. Sawmill workers edged back into the fold, although they still lagged well behind the others, when the issue of extinction was rephrased: "All species have a right to evolve without human interference. If extinction is going to happen, it should happen naturally, not through human actions" (59% of the sawmill workers agreed, and more than three-quarters of the others agreed).

Large majorities of all five groups also agreed with the following two statements: "Humans are ripping up nature, feeling that they can do a better job of managing the earth than the natural system can," and "Humans should recognize they are part of nature and shouldn't try to control or manipulate it." Given these responses, it should come as no surprise that hardly anyone in any of the five groups had much faith in technological fixes. The most optimistic were sawmill workers, but only 15% of them agreed that "we shouldn't be too worried about environmental damage. Technology is developing so fast that, in the future, people will be able to repair most of the environmental damage that has been done."

In this context, it is easy to understand the recent embrace of wildlife. No wild animal more stunningly reflects this turnabout in public perceptions than the wolf. In almost the wink of an eye, the once-loathed creature has become nearly totemic, a representation of all that is wonderful in wildness. In a survey commissioned by Defenders of Wildlife to test public attitudes toward the proposal to reintroduce wolves to the Adirondack Park, the rehabilitation of the wolf's reputation is clearly documented. Had Responsive Management (RM), the polling organization Defenders hired, asked New Yorkers the sorts of questions NORC or Kempton asked of their respondents, it is clear that the results would have been very similar. RM discovered high levels of acceptance of the proposal to reintroduce wolves to the Adirondack Park among New Yorkers from almost all walks of life. Of course, some were less enthusiastic than others; New York has its equivalent of the sawmill workers in the Kempton study. But the impressive thing, again, is how wide support for wolves is; for example, two-thirds of hunters supported wolf reintroduction.[10] As importantly, support for wolves was not simply urban romanticism; although support rose as respondents' distance from the park grew, three-quarters of park residents, the folks who will have wolves for neighbors should the project get the green light, at least "moderately supported" reintroducing wolves to the park.[11]

The most frequently mentioned reason for favoring wolf reintroduction should also come as no surprise, given what we've seen about Americans' attitudes toward nature: 85% of park residents who supported the plan said they were in favor because wolves play an important role in the "balance of nature." We are face-to-face with the belief that had we not driven wolves out in the first place, things would be better now and that if we can bring wolves back, we will be taking a step toward restoring the peace and harmony of a nature in balance.

All the King's Men: Putting Humpty Together Again

As touching as this high-mindedness is, some very bothersome issues—scientific, cultural, and political—are concealed behind the veil of environmental right-thinking. I am not here criticizing wolf reintroduction—far from it. I am worried that the prevailing and broadly accepted arguments for wolf recovery might have the ironic consequence of discrediting even further the ideal of environmental stewardship. But before we explore this worry, let us dig a little more deeply into the reasons for wanting to reintroduce wolves in general and to the Adirondacks in particular.

Reintroducing species to at least parts of their former range and trying to bolster species whose numbers are shrinking can be justified on a number of grounds, not the least important of which is that such efforts, even when they are not as successful as one would hope, nonetheless help counter the gloom and fatalism that so often accompany the narrative of environmental loss.[12] It is nice to be able do something inspiring once in a while (McKibben 1995). More importantly, if being concerned with the environment is always associated with loss and abyss, many will turn away from concern precisely because they'd rather not be depressed all the time. Optimism and success stories are typically far better motivators, in education as well as in mobilizing public opinion, than endless tales of impending doom.

But more than wanting to feel good is involved. At the core is not just the desire to stop or even reverse the decline of biodiversity but the conviction that reintroductions such as that of the wolf in Yellowstone or the Adirondacks will help begin to put Humpty back together. In popular vernacular, it will restore balance to nature, as the majority of the people RM interviewed put it. Experts in biology, ecology, and related fields no longer speak the language of "balance of nature." That metaphor has largely been discredited and in its place we much more commonly read about disturbance, chaos, and complexity in natural systems. As balance fades away, it has been replaced by new metaphors that invite us to continue to imagine a world less likely to spin out of control. One such metaphor is "intact ecosystem."

Thomas McNamee (1997, 114), writer, former president of the Greater Yellowstone Coalition, and ardent supporter of the reintroduction of wolves to Yellowstone, writes of Yellowstone,

> This is the world's first national park, after all, the ur-site of American conservation. Yellowstone Park is also the geographic, ecological, and spiritual center of the largest remaining essentially intact ecosystem in the temperate zones of the earth—the heart of an 18-million-acre complex of wildlands. . . . It is recognized here that the single most powerful absence from the Greater Yellowstone Ecosystem—what demands the cautionary adverb in "essentially intact"—is the absence of the ecosystem's only missing component, the wolf.

Never mind that McNamee, no doubt meaning no insult, leaves Native Americans out and is curiously silent about all the nonindigenous creatures that are around him now. The point I want to call attention to is the metaphor itself and the mischief it does. Although it is left implicit, it is impossible to come away with anything but a sense that once all of the "components" are present, the Greater Yellowstone Ecosystem will become what it was before we started screwing things up.[13] But just what constitutes intactness? Do a few thousand bison, where once there were hundreds of thousands, constitute intactness? Should we worry about how the millions of visitors to the park each year fit into our notion of an intact ecosystem? Although the phrase "intact ecosystem" suggests a hard biological grounding, what is really being claimed is that adding the wolf to the list of species now found in the park makes the collection more nearly complete; visitors to the park can now be assured that the living diorama has representatives of all the major creatures, except Native Americans, that inhabited the region for the past several thousand years. This is all good, but it is a *cultural* good, not a replication of some imagined pristine ecosystem.

To be sure, adding wolves to the mix has added richness to the biological scene. The sites of wolf kills attract a whole host of lesser predators, each trying its best to share in the bounty, and this new source of nutrition will no doubt affect the population dynamics and interactions between the scavenger communities (Robbins 1997). But this dynamic itself is no newer than the dynamic set in motion by the huge fire that swept across Yellowstone Park in 1988. The wolves' sudden reappearance, like the fire, is a disturbance; like all disturbances, whether they are "natural" or authored by humans, it creates opportunities for some animals (and plants) and disasters for others. Beetles may prosper on the leavings of wolves, but the wolves have had a devastating impact on the park's coyotes (Robbins 1997). The sands are continually shifting: Populations rise and fall, species come and go, some to return and others gone forever. Some of these changes take place over long stretches of time, and others happen in the wink of an eye. The point is that with all this flux, notions of intactness, like the earlier notions of climax and natural balance, are deeply misleading.

I am not so much worried about this in terms of how ecologists, naturalists, biologists, and other professionals speak to one another. Scientists, though plagued by the same frailties with which the rest of us contend, nonetheless have well-established means for weeding out good and bad theories. However, the general public is not bound by the discipline of refereed journals and peer reviews. As a matter of fact, the public tends to be skeptical of science, preferring faith and feelings and, we can now add, a belief in a self-healing nature that, given the chance, will produce stability and balance. From a combination of sources, including the most notorious Disney portrayals of wildlife,[14] but also from the language of scientists themselves, the public has been led to believe that if we can set aside habitat and keep our hands off, everything will work out for the best: Wild animals can take care of themselves if we leave them and their habitat alone. This view of nature and our relationship to it casts a thick veil of suspicion over all efforts to manage natural resources except those that can be cloaked in restorationist garb. Benign neglect replaces stewardship and relieves us of all the sticky moral problems attendant on active management of natural resources. Like curators of a museum, we are responsible for ensuring that the collection is not misused or contaminated by outside influences (i.e., people).

Such beliefs as these, played out in the context of the dominant narrative of loss, have spurred efforts to restore and preserve habitats and to reintroduce indigenous flora and fauna to habitats from which we have driven them.[15] Although no one believes that these efforts will undo all the damage we have inflicted on the natural world, setting aside as much land as possible will at least ensure that some semblance of wildness will remain to inspire and instruct us. And so we see earnest efforts to return rivers to their former free-running state and plans to significantly increase the numbers and range of the bison across the midsection of the country. Groups of environmentalists and local activists are busy trying to galvanize support for declaring the Northern Forest that stretches from Maine to upstate New York a national park, and groups are pressing for the reintroduction of wolves, the recovery of Atlantic salmon, and so on. What links these diverse efforts is the belief that with habitat preserved and key ecosystems returned to intactness, all we need to do is learn to live harmoniously with nature, become just one more voice in the symphony of natural harmonies.[16]

On the Wild Side

The desire to live in harmony with nature is not new, but its resurgence in the late twentieth century carries new freight. In part because we have done some protecting of habitat, and perhaps in larger part because we have heedlessly fragmented the landscape, we now face chronic clashes with resurgent wildlife populations, especially those that thrive in the wake of disturbance. We have unwittingly created excellent white-tailed deer habitat all across the country. As a result, from Long Island to Berkeley, California, deer are making their presence felt. Communities long known for tranquility and neighborliness have become bitterly divided over the deer who are eating gardens, ruining landscaping, and creating hazards for drivers. To shoot or not to shoot, that is the question (Dizard 1999; Nelson 1997; Shuey 1997; Kuznik 1998). At least that is the question that divides neighbors. Scientists also are divided on the matter. Some argue that no permanent harm will be done and that, sooner or later, deer populations will decline steeply and herbaceous life will rebound. Others are not so sanguine; they fear that deer will do lasting damage, in effect permanently reducing the diversity of plant life and, as a result, altering the dynamics of the affected area just as surely, though not in the same ways, as would putting in another subdivision. From this angle, the question is less about shooting deer than it is about what kind of nature we want to preserve.

The problem is clear: We have created and protected habitat by design as well as by accident. Good habitat is, by definition, an incubator: The better the habitat, the higher the rates of regeneration and reproduction that go on within it. Right now, for example, Yellowstone Park is excellent wolf habitat: There is plenty of food (elk), wolves face little competition (elk hunting is forbidden; cougars and grizzlies are few), and they have only an occasional rifle to worry about. As a result, the wolf population is expanding robustly. But this is only half the story. The other half is driven not by habitat or food but by the nature of wolves: They are territorial. As big as Yellowstone is and as much wolf food as it contains, wolves can tolerate only so many other wolves in a given area. Good habitat means vigorous reproduction, which quickly sets wolves to dispersing. In broad strokes, the same is true for most species: Robust reproduction means dispersal.

When the globe was "empty" (by contemporary standards), dispersal was an important engine of evolution. An animal would leave a home for which it was well adapted and head for the territories, where it would sink or swim, its fate in part a function of its capacity to adapt to newly encountered conditions. Some made it, many did not. But now things are vastly different: Good habitat is not surrounded by empty spaces.[17] Moose are dispersing from good habitat in northern New England into suburban southern New England, which, despite all the moose food that is around, is not good moose habitat: There are too many commuters, to name only one of the obvious negatives.

Yellowstone is good for wolves (and wolves are undoubtedly good for Yellowstone, but that's another story), which means that adolescent wolves (who will be abundant because the park is good for wolves) will be sent packing. Where will they go? Your guess is as good as mine, but one thing is certain: They will be forced to migrate to less ideal habitat. Inevitably, dispersing wolves will be howling on the urban fringes of Denver, Missoula, and Cheyenne.

Good habitat, to put this slightly differently, is not like a zoo or a museum. It is not bounded. If it is bounded, in the hard sense, it will not remain good habitat for long because, with dispersal opportunities precluded, the populations will destroy what had sustained them.[18]

And there's the deep paradox. Defining environmentalism as preservation means that we are creating reserves that cannot be self-sustaining; worse, the reserves will send forth wildlife with which we cannot avoid conflict. Wolves in Yellowstone are thrilling; wolves in Denver are something else.[19]

Of course, we could trap, shoot, or poison wild animals who show up in places we deem inappropriate, except that growing numbers of Americans reject managing wildlife in this fashion. Having pushed many animals over the brink and many others near the brink, Americans are now inclined to cheer the return of the wild.[20] This change of heart would be fine were it coupled with an acceptance of the fact that resurgent wildlife species make management more, not less, important. This is so not simply for our own safety or convenience but also for the well-being of the animals themselves. Unwittingly, we are creating conditions that will almost certainly result in killing large numbers of wild animals. The reasons as well as the means of killing might well lead us to replace our recently acquired solicitude with a callousness and disregard that will send many species heading back toward the brink. Consider a case that is less well publicized than the white-tailed deer or the wolf: the return of the beaver to New England.

Beavers went extinct all across New England by late in the eighteenth century, which meant that all the roads, sewer systems, parks, and construction for the past two centuries were built with no thought of beavers. Two hundred years later, with trees having replaced the extensively cleared land of a now collapsed agricultural society, the beavers have begun to reclaim their turf. The problem is not simply that they fell trees,[21] nor that they create wetlands, both of which make the landscape hospitable to a more diverse biotic community. The problem is flooding of roadways, basements, septic systems, and other amenities such as playing fields. Water supplies are also placed at risk from *Giardia*, a parasite for which the beaver is a vector.

In my neighboring town, Northampton, Massachusetts, a controversy over beavers has been swirling for 2 years. In 1996, the voters in Massachusetts passed a referendum banning the use of the most effective and efficient traps except when, after a lengthy and convoluted review involving several local and state agencies and boards, it is determined that public health or safety is at risk. A beaver colony that some estimate is home to as many as 50 beavers but is more likely to contain no more than 20, according to beaver specialists, has been causing headaches for town road and sewer maintenance for a number of years, but the few local trappers helped keep the problem from becoming more than a headache. With the trapping ban, the beaver colony grew and the water level rose beyond the headache level. People living near the dam saw their backyards submerged. And the trees that had given residents a sense of privacy as well as a comforting sense of being "close to nature" disappeared.

When the city first proposed hiring a former trapper (who had subsequently become a licensed nuisance animal controller and gone into business with an exterminator), a flurry of protests began. Animal advocates issued impassioned pleas on behalf of the beavers, and letters poured into the local newspaper. The following excerpt from one letter captures the flavor of the probeaver faction. "Beavers mate for life. . . . These are sensitive, reasoning animals who mourn for lost family members, play games, and even appear to show laughter and joy" (Sterste 1998, 10). Another struck a broader theme: "We have more to fear from man and his meddling with the balance of nature than from our fellow animals" (Meyers 1997, B3).

These themes should by now be familiar: Nature is kind, humans should learn from animals and love them, and we should all learn to live peaceably together. Another letter writer, a wildlife biologist and expert on beavers who works for the state, recently advised that we should accommodate beavers by not building near their habitat, advice that ought to have been offered 200 years ago but now seems too little and far too late. Should we buy out homeowners, abandon roads and rail lines, reengineer water supplies and sewer systems, and otherwise adjust our living arrangements to live with however many beavers there happen to be? Who should bear the cost? The letter writer did not even mention such nagging questions, nor did he note that for millennia humans carried out an annual harvest that supplied them with meat and fur and checked the growth of beaver populations. That is, until recently, living with wildlife meant controlling their numbers in one fashion or another. Now, living with wildlife means putting up with the problems wild animals create. How long will we be patient?

The jury is still out. Communities are muddling along with efforts to cope with deer, beavers, geese, and coyotes, to mention only the most common subjects of controversy. State fish and game agencies are scrambling to figure out how best to manage growing wildlife populations in a political context that makes their traditional management practices (trapping and hunting) more and more problematic. One thing does seem certain, though: Wildlife policies are going to become more intensely politicized than ever before. One immediate consequence of this will be to drastically reduce the freedom of action of those most immediately affected by a wildlife problem. State and federal agencies and well-financed national environmental and animal advocacy groups will become embroiled in local disputes and in all likelihood will not be a calming presence. It is also likely that this politicization will not greatly benefit the animals with whom we are trying to live.

Given this context, efforts to reintroduce wolves into the Adirondack Park should proceed with great care. If the reintroduction is successful, wolf dispersal, even though the park is very large, will mean wolves will move toward heavily settled areas where there is bound to be conflict and where wolves are certain to come to grief. Locals will find themselves unable to fashion their own policy responses, which will inflame suspicions of "outsiders," usually government agencies and environmental organizations, and deepen local schisms (Scarce 1998; Wilson 1997). Stakeholders who might otherwise find common ground will be polarized into two camps: one that recognizes the need to control animal populations and one that rejects intervention, especially if it entails killing some animals.

This polarization was prefigured right after the Defenders of Wildlife announced that it was urging a study to assess the feasibility of reintroducing wolves to the Adirondacks. Days after the news release, a writer to our local paper in western Massachusetts deplored the proposal, citing all the usual misinformation about wolves. This sparked several letters sympathetic to wolves. One writer, without a trace of irony, called wolves "gentle predators." Another wrote, "Wolves have a family structure we should envy and emulate"(Alworth 1997, 8). Of course, it is easy to dismiss both the demonization and idealization as uninformed prejudice and naivete, but dismissal misses the point. Like it or not, these will be the terms of the debate, and the result will be a huge erosion of support for stewardship.

My point is simple: Those who would promote wildlife had best make clear that creating good wildlife habitat and reintroducing species does not end our responsibility. In fact, it increases our responsibility and increases the need for management. Setting aside habitat will not produce harmonious nature any more than introducing wolves will restore balance to an area that wasn't balanced to begin with. The return of the wolf promises to increase our love of the wild. If the wolf also helps us recapture a sense of responsibility and stewardship, we will owe it a great deal. If we duck our responsibility, we will have succeeded only in victimizing the wolf yet again, this time with good intentions. I doubt the wolf will care about our intentions.

Acknowledgment

Portions of this chapter first appeared in the epilogue to the revised edition of my book *Going Wild: Hunting, Animal Rights, and the Contested Meaning of Nature* (1999).

Endnotes

1. Burroughs was by no means an apologist for exploiters of nature any more than he was indifferent to the damage industrialism and urbanism were inflicting on the landscape. Indeed, he was one of the more persuasive of the mounting chorus of voices demanding that our natural resources be protected from heedless overexploitation. He was closely aligned with Theodore Roosevelt's and Gifford Pinchot's efforts to institutionalize the ethos of stewardship in our nation's use of natural resources.

2. Mitchell has recently written an appreciative introduction to a slender but rich volume in this genre by Ellsworth Barnard, an emeritus professor of English at the University of Massachusetts. Barnard (1998) takes his reader on an excursion to the place in which he grew up, now a Massachusetts Audubon Society sanctuary. Mitchell correctly locates Barnard's evocation of place alongside Burroughs but, I think, mistakenly includes Muir as well. Barnard, like Burroughs, sought stimulation at home, more like Thoreau's "wide travels" in Concord than like Muir's questing. The distinction may seem small, even petty, but the difference is large. Barnard, Burroughs, and Thoreau accepted human presence in the landscape and even human modification of the landscape (though not all human impacts). Muir sought to escape the landscapes that had been modified by humans. Wildness for Muir was a very different proposition than it was for Thoreau and, for that matter, than it is for Mitchell.

3. For their pains, Budiansky and Easterbrook have been roundly excoriated in the environmental press, a press that seems resolutely committed to juxtaposing human-made travesties to unspoiled natural beauty as though there can be nothing of worth in between. Environmental historian William Cronon has experienced a similar rebuke for having the temerity to suggest that environmentalists' commitment to preserving the pristine has inadvertently had the effect of tilting environmental concerns away from the more prosaic but nonetheless essential need for stewardship of even the most disturbed and modified landscapes (Cronon 1995a). For a thoughtful critique, see Waller (1998).

4. Moose are now commonly seen in central and western Massachusetts and occasionally are seen in eastern Massachusetts and further south. In 1998, a driver collided with a moose in Connecticut.

5. Of course, wildlife species are not the only beneficiaries of the Endangered Species Act. Plants also have become protected. A full account of the species involved is beyond the scope of this chapter.

6. In a recent essay, David Quammen (1998) deftly skirts doom and gloom as he invites us to come to terms with what the current wave of extinctions will mean. Although it might not mean "the end," it will almost surely mean a much less rich and varied world. We have no way to imagine what, if anything, will fill the niches vacated by the species we are displacing.

7. Unfortunately, the worriers seem to be concentrated in Rep. Chenowith's electoral district, thus ensuring her seat in Congress, from which position she has been able to make much mischief.

8. In fact, Thoreau's real wilderness experience in the deep woods of Maine left him

anything but uplifted. In short, he hated it. His friend Emerson, whose nature writings were even more metaphorical than Thoreau's, was similarly uncomfortable in the great outdoors. Muir records how disappointed he was when, after he'd arranged a camping trip in the redwoods of northern California for himself and Emerson, Emerson stayed at the site for a few hours and then asked to be taken to a hotel.

9. The samples from each group were not random, so generalizing beyond the participants themselves is not warranted. Still, their results certainly corroborate the general orientation being sketched here.

10. Unfortunately, the RM research report does not offer data on the one group that might be most unhappy with wolves: farmers. Every indication is that farmers would hold the most traditional (i.e., negative) views of wolves, although this may reflect more the views of the organized voices of farmers (e.g., the Grange, Farm Bureau) than of individual farmers.

11. To be sure, there are questions about the accuracy of the results RM obtained. Local officials of the communities located in the park do not think the poll reflected the sentiments they have been hearing from constituents. Indeed, the organization of elected officials in the park has overwhelmingly voted to oppose any plan to reintroduce wolves to the park. But it is clear that this opposition does not arise from a wolf phobia so much as a deep-seated mistrust of outsiders and their presumed penchant for dictating to locals how they should live. I dare say if the choice were between living with wolves or living with state or federal policymakers, the locals would choose wolves in a heartbeat.

12. We could go into much greater detail than space here permits. In addition to gloom and fatalism, the narrative of loss also contains a large dose of misanthropy. To cite but one example of this, in the Kempton et al. study people were asked whether they agreed with the following statements: "If any species has to become extinct as a result of human activities, it should be the human species" and "I would rather see a few humans suffer or even be killed than to see human environmental damage cause an entire species go extinct." EarthFirst!ers led the pack by a large margin: 80% agreed with the first statement and 90% with the second. Only the dry cleaners, among the other four groups polled, came anywhere near this mark: 56% agreed with the second statement.

13. One of the more heroic efforts based just on this premise is appreciatively reported by Donovan Webster (1999). Ted Turner, as Webster details, has decided to do all that he can to restore his holdings in the West to their "original condition." Miles of fencing have been removed. Exotic plants are relentlessly removed. Bison have replaced cows. If nothing else, the effort promises to teach us a great deal about what we've lost for good and what we can recoup. But it also makes plain how such efforts, whether underwritten by personal fortune or government appropriation, are experiments rather than blueprints.

14. Disney's transformation of *Bambi* from an Austrian socialist allegory to a bourgeois morality tale is the most often cited example of how nature has been idealized beyond recognition in popular culture. The latest of Disney's efforts to "educate" Americans about nature may even be worse. Disney's new Animal Kingdom theme park in Orlando purports to be naturalistic, even though its presentation of

wild animals is every bit as unreal as the cartoons for which Disney is famous. For a scathing review of Animal Kingdom, see Tad Friend's "Please Don't Oil the Animatronic Warthog" (1998).

15. As in most reform movements, good intentions and zeal combine to make polite fictions acceptable. Thus, the successful effort to establish breeding pairs of eagles near large reservoirs has been billed as a reintroduction, even though eagles would have found nothing of interest to them in these areas before the reservoirs were built. Similarly, the California condor has recently been "reintroduced" into a remote area of Arizona and New Mexico, despite very shaky evidence that the birds had ever lived there. Intense debates also arise over the genetic strains of the animals being reintroduced: Are these the real McCoy or something more nearly exotic? I will leave a discussion of the moral career of the term *exotic* for another day.

16. To be sure, there is a harder edge. Some elements of contemporary environmentalism have become so captured by the narrative of loss that they have turned misanthropic. Saving whales, wolves, spotted owls, or redwoods becomes not just a good thing to do but the key to righting the wrongs of the world. Just as the "right-to-life" movement has its extremists who, in the name of life, are prepared to kill, the environmental and animal rights movements have their extremists.

17. No one has made this point more emphatically than David Quammen (1998).

18. Isle Royale is a good example of this. Over the past five decades moose have periodically overbrowsed the island, and their numbers dropped precipitously as a result. Wolves fared no better. Their numbers rise and fall, less a function of the availability of moose than a function of inbreeding and conspecific strife (Botkin 1990).

19. Lest this seem excessive, dispersing wolves have been known to travel up to 500 miles from their natal homes. More importantly, the only "naturally occurring" wolf population in the lower 48, the wolves in Minnesota, has steadily expanded its range since the Endangered Species Act protected them in the early 1970s. Since then, their numbers have more than tripled, and dispersal has led to well-established populations in northern Michigan and northern Wisconsin and breeding pairs of wolves resident 60 or so miles north of Minneapolis–St. Paul. David Mech, the preeminent expert on wolves, estimates that it will soon be necessary to kill 300 to 500 wolves a year in Minnesota to retard dispersal and thus minimize the chances that wolves will inflame urban hostility.

20. By this I do not mean to imply that Americans have become uniquely solicitous, although Kellert's (1994) research suggests that solicitude for wildlife does not extend much beyond the reach of Western culture.

21. Felling trees is not a general problem; New England is far more heavily forested than it was a hundred years ago. But in specific locales, beavers have decimated the trees that have buffered residential developments. There is no small irony in this. Dense housing developments that cluster dwellings tightly so that open space can be preserved are touted as environment-friendly. But these developments typically depend on trees to screen neighbors. When beavers move in, neighbors suddenly discover they are closer to one another than they had bargained for.

Part III

Precedents and Antecedents in Law and Regulation

In this part, two chapters examine the ways in which Anglo-American law bears on the issue of environmental restoration.

In Chapter Nine, "Tracking a New Relationship: The Wolf in New York Law," Nicholas Robinson, the Gilbert and Sarah Kerlin Distinguished Professor in Environmental Law at Pace University School of Law and former deputy commissioner and general counsel of the New York State Department of Environmental Conservation, explains the role that New York law plays with regard to wolves.

In the first part of his chapter, Robinson illustrates how law reflects particular cultural and social commitments. Rejecting British law that reserved forest hunting to the aristocracy and designated wolves off-limits as "beasts of the forest," early Colonial legislatures in America encouraged destruction of wolves by adopting laws providing bounties for each wolf killed. By the mid-nineteenth century, bounty laws were repealed principally to conserve tax dollars that were spent compensating bounty hunters. In addition, by that time, wolves were rare if not gone from New York.

After World War II, the expanded role of the science of ecology began to change public conceptualizations of nature. In 1969, New York's legislature enacted the Mason Law to prohibit trade in specified endangered species, expressly including in the law the gray wolf *(Canis lupus)* and the red wolf *(Canis rufus)*. To this the legislature added the Harris Law to authorize New York to promulgate its own lists of endangered species independently of the federal government lists. New York's authority to protect the wolf, even though it was no longer indigenous to the state, was upheld in the courts. Henceforth, no one in New York could trade in wolf furs or parts, and the market in wolf parts was wholly forgotten. These New York laws were complemented by the Endangered Species Act of 1973 at the federal level, which inaugurated new prohibitions on the taking of species that were threatened with extinction.

Consideration of wolf reintroduction today is governed by the New York State Environmental Quality Review Act (SEQRA), which is the environmental impact assessment law for the state inspired by the National Environmental Policy Act (NEPA). The purpose of SEQRA is, among other things, to "encourage productive and enjoyable harmony between people and the environment." Robinson outlines the steps required under SEQRA for considering the biological, social, economic, and ethical issues pertinent to wolf reintroduction.

By compiling the written record of the environmental impact statement, Robinson says, New Yorkers will be weighing in on the meaning of the environment and its relationship to wolves in New York. Is the environment the set of ecosystems in New York that have evolved for 200 years without wolves, or is it the ecosystem that evolved over the earlier millennia with wolves? Is it an environmental impact to introduce a wolf into a world with roads and shopping centers? Could wolves have survived in New York as the human population grew and land development became common? Even if introduced, will future changes, such as global warming, affect the ecological viability of wolves in New York? Although New York's Adirondack Mountains are the largest wilderness area east of the Mississippi River, are its "wild" habitats sufficient to sustain a viable wolf population?

Legal questions also are implicated in this environmental impact assessment. Because New York law has secured property owners and the residents of towns from wolves for more than two centuries, if wolves were to be introduced, what new safeguards should the law provide to property owners in lieu of the wolf eradication and bounty programs of the past? What are the appropriate ways to mitigate any adverse effects on economic or recreational interests caused by the presence of wolves? Because the Forest Preserve has been provided for in the New York State Constitution since 1894 and wolves were not a part of this "wild forest land" at the time of ratification, how, as a matter of constitutional law, should "wild" be understood: with or without wolves? Also, because the State Department of Environmental Conservation administers the Forest Preserve but the Adirondack Park Agency administers the land use and development aspects of the vast private lands of the Adirondacks, how will these two agencies view the question of environmental impacts associated with wolf reintroduction?

Robinson refers to recent New York law regarding wetlands to anticipate some of the answers to these questions. In a period of little more than 3 years, he says, New York laws shifted from regarding wetlands as a nuisance to safeguarding the remaining wetlands as natural treasures serving society. The owner of wetlands who once could despoil them with the blessing of the state is now a steward of these services.

In Chapter Ten, "Legal and Policy Challenges of Environmental Restoration," Joseph Sax, deputy assistant secretary for policy at the U.S. Department of the Interior under the Clinton Administration and James H. House and Hiram H. Hurd Professor of Environmental Regulation at the University of California, Berkeley, explains the dramatic implications of restoration and stewardship for the legal system's agenda for land.

According to Sax, the conventional legal understanding is that private land is there to be used for economically productive purposes: agriculture, stock-raising, residences, and industrial and commercial purposes. To introduce programs that intrude on these purposes is, by ordinary proprietary standards, a diversion from land's "productive use" mission. The wolf may be only a small diversion, but in the conventional perspective it is symptomatic of something quite drastic and disconcerting. To make everyone accommodate the wolf, rather than the other way around, is to turn things upside down, to put in question the very purpose of the private property system in land.

Drawing on precedents set in historic landmark preservation, from Stonehenge to Grand Central Terminal in New York City, Sax explores the very modern notion of a public heritage and the ways in which it might be used to preserve our biological and cultural capital.

Paradoxically, Sax observes, the same forces that gave rise to concerns about saving a world that seemed to be succumbing to voracious industrialization also generated ideas of progress, conquest, and inevitability that threatened to sweep aside everything in their path. People during the last two centuries have lived largely by an ideology that required and even idealized a transformational economy that ensured that future generations would live in a world nothing like that which had gone before. Forests were turned into fields for the plow; wildlife was exterminated in favor of domestic herds; rivers were dammed for hydropower first for mills, then for irrigation, electricity, and cities; prairies were extirpated; and wetlands were drained and filled to support agriculture.

Citing the abandonment of slave ownership in the United States and the laws of primogeniture and a husband's property rights in his wife's estate, Sax argues that changing social values and goals have undeniably resulted in changes in the conception of property. Where "natural capital"—that is, biodiversity and healthy ecosystems—is concerned, Sax argues, we can neither continue on the path of the old transformational economy nor return to an economy of nature that reflects the world in some uninhabited form or in the form of habitation by a hunting and gathering society or any other form of nonindustrialized (or postindustrialized) world. The challenge is to decide what sort of adaptations of conventional development will permit land and water to serve agricultural, industrial, recreational, and municipal needs while sustaining a more desired level of biological diversity.

Chapter Nine

Tracking a New Relationship: The Wolf in New York Law

Nicholas A. Robinson

Cultures imagine nature through myriad means, among them creation myths, economic practices, social traditions, and laws. Because of the clarity needed for prescriptive legislation and the precision essential to a judicial decision's resolution of a dispute, laws provide fairly explicit insights into how a social order regards an aspect of nature. So it is with wolves. Laws legitimized the human war on wolves for centuries. Since the 1970s, laws have sanctioned the protection of wolves and the restoration of their populations. In New York, the state's environmental impact assessment laws will provide the primary forum for deciding how and whether wolves may be reintroduced into natural settings in the state.

Although federal laws, such as the Endangered Species Act and the National Environmental Policy Act, will play a role in any effort to resettle wolf populations in New York (because the Fish and Wildlife Service will need to approve the movement of a wolf, as an endangered or threatened species, from Canada or another state into New York), federal involvement probably will proceed only after New York has decided to allow the wolf repatriation. Therefore, it is New York State law that will be the determining factor. There are almost no federal public lands in New York, and the state itself has established the vast Adirondack and Catskill Forest Preserve, forever wild forest land that might be suitable for wolves. More importantly, it is at the state and local levels that the new social consensus about wolves, as either a threat or a benefit to nature and to human society, will be forged, providing a basis for any new legal status for wolves in New York.

Laws Banishing Wolves

The era of legally sanctioned wolf extirpation predates the colonization of New York. *Manwood's Treatise of the Law of the Forest,* published near the end of the reign of Queen Elizabeth I, restated the law of Britain concerning wolves as

"Beasts of the Forest" as opposed to "Beasts of the Chase." The difference was hardly academic, for so long as the Beasts of the Forest stayed in the forest, they were "privileged" and "were not to be hurt there." This was no example of an early concern for a species even then threatened with extinction throughout Britain. *Manwood's Treatise* explained the reason: "As for Wolves, there are none now in England, but formerly there were many here; and all the ancient Woodmen and Foresters did account them Beasts of the Forests, and the Fox a Beast of the Chase, because Kings and Princes took a Delight in hunting them; and therefore, if they were killed by any other Person without Leave, and within the Boundaries of the Forest, it was a Breach of the King's Royal Franchise, and the offender was to be punished. But the Wolves were almost destroyed by King Edgar, so that when Canutus began to reign, no care was taken to preserve those which were left, because they were dangerous and hurtful Beasts" (Nelson 1717, 160–161). In Britain, hunting wolves was a royal sport, at least in the Royal Forest, and the public was barred from the endeavor.

Although English law was received by the Colonies and continued to be the law of New York by act of the New York legislature after the state became independent, the early immigrants chose to disregard the aristocratic privileges that had reserved forest hunting to the aristocracy in Britain. There were no royal forests in the New World; all was forest. Anyone could and did harvest wildlife. Because wolves were a threat to domesticated animals, early colonial legislatures encouraged their destruction by adopting laws providing for payment of a bounty for each wolf killed. Such bounty laws remained in place long after wolves were no longer a threat[1] and were emblematic of the received wisdom of society that each wolf was an enemy or, as Theodore Roosevelt, the first president to call for nature conservation laws, expressed it in his essay "Wilderness Hunter," the wolf is "the archetype of raven, the beast of waste and desolation" (Roosevelt 1900, 46).

From the early seventeenth century onward, the laws of New York, like those of the other colonies and states, settled the practice to establish bounties for the killing of wolves (Cronon 1983). Even when the threat of wolves was not apparent, the presence of bounty laws stimulated bounty hunting as a ready means for hunters to earn money, and wolves were tracked and killed. Bounty hunters would track a wolf far afield of the township that offered the bounty, and then bring the dead wolf to the local government that offered the highest sum. As Thomas Lund notes, "Since few legislators could long survive the use of local tax funds to import dead wolves, restrictions were soon devised" (Lund 1980, 88). Bounty hunters would bring in the ear of a wolf as proof of the killing and claim a bounty. In New York City, so many wolf ears appeared that the city came to require hunters to produce the entire head, with both ears attached, to avoid being obliged to pay the bounty twice (Lund 1980).[2]

By the mid-nineteenth century, wolves were rare if not nonexistent in New York and most of the northeast United States.[3] It appears to have been a desire to conserve tax dollars, rather than any change in attitude toward wolves, that prompted repeal of bounty provisions. Indeed, in an echo of such attitudes about predators as threats to agriculture or other social values, in 1955 New York enacted authority for town boards of supervisors to pay bounties for the killing of foxes.[4]

From the end of the nineteenth century until World War II, the laws and policies among other states and among federal agencies continued to promote the elimination of wolves (Mighetto 1991). As a result, by 1973, wolves had been eradicated, through the bounty system or the efficient use of poisons, from most of their habitat in the eastern states, the Great Plains, and the western public lands. The only remaining wolf populations in the United States outside Alaska were in the northern part of Minnesota, on Michigan's Isle Royale, and along the Canadian border[5] (see Mech, Chapter Two, this volume; Peterson, Chapter Twelve, this volume).

Legislation Rethinking Human Conduct Toward Wolves

After World War II, the expanded role of the science of ecology began to change public conceptualizations of nature. By the time Earth Year captured the imagination of New Yorkers in 1969, attitudes toward predators, wildlife, and endangered species in particular had begun to change. New York's legislature enacted the Mason Law to prohibit trade in specified endangered species, expressly including in the law the gray wolf *(Canis lupus)* and the red wolf *(Canis rufus)*.[6] To this the legislature added the Harris Law to authorize New York to promulgate its own lists of endangered species independently of the federal government lists.[7] New York's authority to protect the wolf, even though it was no longer indigenous to the state, was upheld in the courts.[8] Henceforth, no one in New York could trade in wolf furs or parts, and the market in wolf ears was wholly forgotten. These New York laws were complemented by the Endangered Species Act of 1973[9] at the federal level, which inaugurated new prohibitions on the taking of species that were threatened with extinction. It is interesting to note, however, that the earlier presumption against wolves as noxious beasts also persisted in New York's laws during this time. In northern New York, the Environmental Conservation Law in 1972 continued the provision authorizing people "in wolf or coyote hunts with dogs" to carry whatever firearms they needed as long as they had a permit from the state conservation officer in the locality.[10] Because no wolves remained in New York, this provision was reenacted as an atavistic anomaly, but it illustrates the power of the idea that wolves should be eradicated.

Ironically, by the mid-1970s, when New York law would have afforded protection to wolves, they had been extirpated from this part of their range. Today, because of natural and human-made barriers, wolves have little prospect of reentering this range in a natural migration. The Great Lakes and human settlement and industrialization separate New York from the wolves now in Minnesota, Michigan, and Wisconsin. Likewise, the breaking of winter ice to permit year-round navigation on the St. Lawrence Seaway eliminates the only corridor by which wolves might reenter New York on their own from Canada.

The only way wolves may be restored to their historic habitat in New York, therefore, is for humans to reintroduce the species. There is a significant body of scientific and public policy literature about the management of wolves on Isle Royale (see Peterson, Chapter Twelve, this volume), the Boundary Waters Canoe Area, or the Upper Peninsula of Michigan (Dunlap 1988; Madonna 1995) and about wolf reintroduction into Yellowstone National Park (see Clark and Gillesberg, Chapter Eleven, this volume; Wolok 1996; Chase 1987; Kluger 1998) and the 1998–1999 release of Mexican gray wolves into the Apache–Sitgreaves National Forest in Arizona (Slater 1999). Suffice it to say that the experience elsewhere is that any change in policy about wolves in these other regions of the United States has been accompanied by tremendous public controversy. The traditional fears about the wolf inspire robust opposition to programs to expand the numbers of wolves or to restore them to their historic range. Such controversies accompanying wolf management suggest that any decision to introduce wolves back into New York will be highly visible and contested.

Although the New York state legislature could enact a law governing the repatriation of wolves to New York (as it did when it banned the sale of wolf fur and classified the wolf as an endangered species under state law), there is little evidence that a grass-roots political movement in favor of any such new statutes exists. Moreover, because existing law would permit wolf reintroduction, legislators will find that there is no need to take up the issue unless the State Department of Environmental Conservation makes a decision pro or con and constituents seek a law to reverse the department's decisions after challenging them through judicial review or some other court proceeding. Because very different convictions about nature and wolves will be in contention with respect to any proposal to reintroduce wolves in New York, the legal procedure through which the decision will be made is as important as the decision itself. Indeed, how the process develops will largely influence the outcome and the decision.

Legal Procedures for Determining Wolf Restoration

In New York, the law governing the process for such a decision is the State Environmental Quality Review Act (SEQRA),[11] which is the environmental impact assessment law for the state, inspired by the National Environmental Policy Act (NEPA).[12] The State of New York has assigned responsibility for managing wildlife to the Department of Environmental Conservation (DEC)[13] and has expressly required that any person seeking to possess, transport, or import a live wolf must obtain a permit from DEC.[14] Although this provision is designed primarily to regulate people who have wild animals in their possession, when it is read together with the Environmental Conservation Law's requirement that "no person shall willfully liberate within the state any wildlife except under permit from the department,"[15] it is evident that any effort to import a wolf or wolves, and then release them within New York, will require DEC permits. Within the DEC, the Bureau of Wildlife has the lead responsibility to support the commissioner of environmental conservation in whatever decision he or she makes on such permit applications. It is these permit decisions that, in turn, necessitate the use of SEQRA as the procedure for deciding whether to grant or deny any prospective application for permits.

Chapter Three of this volume, by Robert Inslerman, describes the role of the New York DEC in wolf restoration decisions.[16] Inslerman states that any decision to reintroduce wolves to New York would require analysis of the historical record (whether a particular wolf species really was in New York), the biological feasibility (whether the habitat of the Adirondacks can sustain wolves), ecological consequences (the effects of wolves' return on other species), consequences to people (e.g., effects on economy, hunting and trapping, land use, and domestic animals), social acceptability (DEC would not support wolf restoration without the informed consent of the people of the state, particularly those in the affected area), and the costs to DEC (how DEC could cover its own incrementally additional costs for managing introduced wolves).

All of the issues identified by DEC are subsumed within the requirements of SEQRA. Indeed, SEQRA requirements are quite broad, applying to "projects or activities involving the issuance to a person of a lease, permit, license, certificate or other entitlement for use or permission to act by one or more agencies."[17] Likewise, the New York Court of Appeals has ruled that "the question of significance is not arrived at solely by gathering data and making calculations; instead, it is ultimately a policy decision, governed by the rule of reasonableness, that the particular facts and circumstances of a project do or do not call for preparation of a full impact statement."[18] Therefore, there is little doubt that the DEC would need to make a determination that the application for permits to reintroduce wolves into New York is an action significantly affecting the environment and thus requiring full SEQRA procedures, including preparing an environmental impact statement (EIS), preparing a positive declaration on the necessity of doing so, preparing and circulating a draft impact statement, holding a public hearing and receiving written comments, and responding to every substantive comment in a final EIS.

Although the several broad issues raised in Chapter Three are relevant to the reintroduction of wolves into New York, a more specific delineation of these and other issues will be needed. SEQRA requires the DEC to hold a meeting to determine the full scope of the environmental impact assessment study.[19] The DEC ultimately is responsible for preparing the environmental impact statement, but the DEC has a choice of either preparing the EIS itself or asking the applicant for the permits to prepare it. SEQRA provides that "if the applicant does not exercise the option to prepare such statement, the agency shall prepare it, cause it to be prepared, or terminate its review of the proposed action."[20] Even if DEC prepares the EIS, it can charge the costs of the preparation to the applicant,[21] so the applicant will pay for a substantial study for the EIS whether the applicant does the study or whether the DEC does the study. If the applicant prepares the EIS, DEC can charge a reasonable fee for its costs of reviewing the EIS before accepting it as complete.[22] Only a well-financed organization is likely to apply to introduce wolves back into New York, for it will bear the major cost of the scientific and social studies within the EIS.

Applicants often prefer to prepare the draft EIS themselves so that they can ensure that it is competent and complete enough to support a decision in their favor. The public has a right to comment on the draft EIS,[23] as do all other interested agencies. It is in the public comment process that the pros and cons of any wolf reintroduction will become focused and clear. In view of the disputes involving wolf management in other parts of the United States, it is likely that the comments will fall into four categories: general comments by individuals and nongovernment organizations either for or against the reintroduction (environmental advocates, individuals, property owners, and a wide range of views will be expressed for and against wolves, often without supporting documentation), scholarly or professional comments by wildlife experts providing a sort of peer review of the scientific analysis in the draft EIS (in any scientific study, there are always further scientific questions, and the scientific method promotes further study rather than establishing one consensus), economic and other studies by commercial or other vested interests that are apprehensive that wolf reintroduction will impair their interests (the Farm Bureau on behalf of agriculture and its domesticated animals), and local governments or other state agencies commenting from the perspective of their jurisdiction (the New York Association of Towns already has expressed apprehensions about introducing wolves). Given the range of interests that will be concerned about wolves in New York, the DEC will feel obliged to hold one or more public hearings on the draft EIS. The Court of Appeals has observed that "if the agency [DEC] determines that there is sufficient interest and that it would aid decision-making or provide an efficient forum for public comment, the agency should hold a public hearing on notice."[24]

SEQRA requires the DEC to consider and respond to every substantive comment made on the draft EIS.[25] Given the need to ensure that the DEC decision will be independent and be perceived to have been independent and without bias, it is likely that a DEC hearing officer, usually an administrative law judge, would preside at the hearings and prepare the final EIS (Gerrard et al. 1995). The same administrative law judge would prepare the draft findings of fact for the Commission of Environmental Conservation.[26]

Because the importation of wolves as a species listed as endangered under the federal Endangered Species Act will also require a federal permit from the U.S. Fish and Wildlife Service, an applicant could consider applying for that permit before it applies to the DEC. Under SEQRA, where a federal EIS is done pursuant to NEPA, there is no need for an impact assessment under SEQRA.[27]

Under NEPA, the federal government bears the cost of preparing the EIS, so an applicant might prefer to avoid bearing the costs, as it would under SEQRA. However, given the fact that most or all of the impacts associated with wolf reintroduction in New York will be in New York, and given the current strong judicial preference of the U.S. Supreme Court that the sovereignty of the states should be recognized unless there is a clear provision in the U.S. Constitution assigning the authority to the federal government,[28] it may well be that the federal agency in any event will defer to the DEC. The federal agency can then use the state EIS as the basis for its own and tier its reviews on top of the state's.[29]

These SEQRA procedures, then, become the common rules that all interests must follow in the debate over reintroducing wolves into New York. SEQRA obliges applicants to show how they will mitigate any adverse environmental effects of their project. The law does not define precisely what such an adverse effect is or how to mitigate it. The courts require use of a rule of reason: the commissioner of environmental conservation shall determine whether the impacts are reasonable in the circumstances and on the factual presentations in the record assembled by the DEC staff through the SEQRA process.

Thus, for there to be a reasonable SEQRA determination, New York's law requires a thorough deliberation of all the biological, social, economic, and ethical issues pertinent to wolf reintroduction into New York. The other chapters in this book constitute a forecast of many of the issues within the scope of the EIS. The multidisciplinary nature of these issues entails exploration of some very fundamental questions. What is the environment? What is an impact? What stewardship roles should people have toward wildlife? Responses to these basic queries will be assembled within the record of studies, policy arguments, and analysis that make up the EIS under SEQRA. Empirically, this written record will define the parameters of what is reasonable.

Reconceptualizing the Wolf in New York

By compiling the written record of the EIS, New Yorkers will be weighing in on the meaning of the environment and its relationship to wolves in New York. Is the environment the set of ecosystems in New York that have evolved for 200 years without wolves, or is it the ecosystem that evolved over the earlier millennia with wolves? Is it an environmental impact to introduce a wolf into a world with roads and shopping centers? Could wolves have survived in New York as the human population grew and land development became common? Even if introduced, will future changes, such as global warming, affect the ecological viability of wolves in New York (Post et al. 1999)? Although New York's Adirondack Mountains are the largest wilderness area east of the Mississippi River, are its habitats sufficient to sustain a viable wolf population?

Legal questions also are implicated in this environmental study. Because New York law has secured property owners and the residents of towns from wolves for more than two centuries, if wolves were introduced, what new safeguards should the law provide to property owners in lieu of the wolf eradication and bounty programs of the past? What are the appropriate ways to mitigate any adverse effects on economic or recreational interests caused by the presence of wolves? Because the Forest Preserve has been provided for in the New York State Constitution since 1894 and wolves were not a part of this "wild forest land" at the time of ratification, how, as a matter of constitutional law, should *wild* be understood: with or without wolves? Also, because DEC administers the Forest Preserve but the Adirondack Park Agency administers the land use and development aspects of the vast private lands of the Adirondacks, how will these two agencies view the question of environmental impacts associated with wolf reintroduction? Should the Catskills also be a target habitat for wolf restoration?

SEQRA defines *environmental* to mean "physical conditions which will be affected by a proposed action, including land, air, water, minerals, flora, fauna, noise, objects of historic or aesthetic significance, existing patterns of population concentration, distribution, or growth, and existing community or neighborhood character."[30] Is it an aesthetic benefit that the night air of the Adirondacks will be punctuated with a wolf call, or is the quiet of the night as enjoyed by Bob Marshall (Glover 1986) the sort of wilderness experience that a visitor to the Adirondacks should enjoy?

The answer to these questions may be found in the analogy to how New Yorkers came to change their views about wetlands. The tidal estuaries and marshes of New York once were thought of as waste lands that had to be reclaimed by dredging or filling. Until enactment of the Tidal Wetlands Act in 1973,[31] New York's Navigation Law actually mandated the reclamation of the swamplands of New York (see Sax, Chapter Ten, this volume). Ecologists taught how valuable the "undeveloped" marsh was, and after 3 years of heated debate in the assembly and senate, and with Governor Rockefeller, the legislature stated that the "tidal wetlands constitute one of the most vital and productive areas of our nature world" because wetlands naturally provide flood control, hydrologic absorption of storm surges, siltation control, habitat for migratory species (whether ducks and wildfowl or songbirds), the food chains and nurseries of many fish species, and open space and scientific study.[32] In a period lasting little more than 3 years, New York's laws shifted from regarding wetlands as a nuisance to safeguarding the remaining wetlands as natural treasures serving society. The wetland owner who once could despoil them with the blessing of the state is now a steward of these services.

Conclusion

Will the SEQRA procedures facilitate a comparable evolution in thinking about the habitat and lives of wolves in New York and permit New Yorkers to embrace another dimension of Aldo Leopold's "land ethic" (Leopold 1949)? Will the public and private custodians of lands in New York that could biologically support wolves regard it a benefit of ownership to experience wolves migrating across their land? Will they regard wolves as a beneficial force, limiting the number of beavers that now proliferate in New York with near impunity, flooding farm and recreation lands alike? Can the introduction of wolves inspire the state and landowners to develop habitat conservation plans and manage ecosystems across property lines? Will tourism in the Adirondacks or the Catskills be enhanced as visitors seek the thrill of seeing or hearing wildlife?

SEQRA's purpose is to "encourage productive and enjoyable harmony" between people and the environment, to promote measures to "eliminate damage to the environment" and "enhance human and community resources," and ultimately "to enrich the understanding of the ecological systems, natural, human and community resources important to the people of the state."[33] If we cannot know today whether wolves will be accepted back into New York, we can at least know the law governing how New Yorkers will debate and determine whether reintroducing wolves is compatible with their evolving vision of the natural environment.

Endnotes

1. For instance, in 1847 in New Hampshire, the instructions to town officials provided that "if any person shall kill any wolf, or wolf's whelp within this state. And shall produce the head thereof to the selectmen of the town within which it was killed, or if there be no selectmen in such town, then to the selectmen of the nearest town having such officers, and shall prove to the satisfaction of such selectmen

that such wolf, or wolf's whelp was killed by himself, or by some person whose agent he is, the selectmen shall cut off the ears from the head so produced, and shall otherwise disfigure the same that it shall never again be offered for a bounty, and shall pay to such person or his order twenty dollars for every wolf, and ten dollars for every wolf's whelp as aforesaid" (R.S. ch. 127, sec. 1. "A Guide to Officers of Towns." Concord, N.H.: G. Parker Lion, 1847, pp. 367–368).

2. See 1723 New York Laws, ch. 443, 17th Assembly, 4th Session, in 2 Colonial Laws of New York (1894), at pp. 163–164.

3. In the Catskill Mountains of New York wolves were still reported as late as 1833. See Evers (1982, 395). In 1830, William Cullen Bryant wrote that the Catskills were "a wide sylvan wilderness, an asylum for noxious animals, which have been chased from the cultivated regions, the wild cat, the catamount, the wolf and the bear, and a haunt of birds that love not the company of man" (Bryant 1830, 9).

4. New York Laws 1955, ch. 630.

5. In Minnesota the legislature in 1966 enacted a law to continue paying a bounty on wolves, and governor Karl F. Rolvaag vetoed the bill, thus ending the bounty (Dunlap 1988, 147–148).

6. L. 1976, ch. 49. See Sections 11-0535 and 11-0536 of the Environmental Conservation Law of the State of New York, vol. 17-1/2 *McKinney's Consolidated Laws of N.Y.*

7. L. 1980, ch. 800, Environmental Conservation Law Section 11-536(4); see lists at vol. 6 N.Y. Code of Rules and Regulations, Section 182.

8. *A.E. Nettleton Co. v. Diamond,* 27 N.Y. 2d 182, 315 N.Y.S. 2d625, 264 N.E. 2 118 (1970). The state laws also were held not to violate the U.S. Constitution. *Palladio, Inc. v. Diamond,* 321 F. Supp. 630 (SDNY 1970) aff'd 440 F. 2d 1319 (2d Circuit), cert. den. 404 U.S. 983, 92 S. Ct. 446, 30 L. Ed. 367.

9. 16 U.S. Code, Section 1531, 87 Stat. 884.

10. Environmental Conservation Law, Section 1-0931(5)(c); L. 1972, ch. 713.

11. Environmental Conservation Law, Article 8.

12. 42 U.S. Code, Section 4331, et seq.

13. Environmental Conservation Law Section 11-0303 states, "The general purpose of powers affecting fish and wildlife, granted to the Department by the fish and wildlife law [Article 11 of the Environmental Conservation Law], is to vest in the Department, to the extent of the powers so granted, the efficient management of the fish and wildlife resources of the state."

14. Environmental Conservation Law Section 11-0511 states, "No person shall, except under license or permit first obtained from the department, possess, transport, or cause to be transported, imported or exported any live wolf, wolf dog, coyote, coydog, fox, skunk or racoon."

15. Environmental Conservation Law Section 11-0507(3). The section also provides that the DEC "may issue such permit in its discretion, fix the terms thereof and revoke it at pleasure."

16. See also variants of this chapter by Hicks et al. (1996) and Inslerman (1998).

17. Environmental Conservation Law Section 8-0105(4)(I); the definition of the permit is the regulations for SEQRA at vol. 6 N.Y. Codes of Rules and Regulations, Section 617.2(aa).

18. *Coca-Cola Bottling Co. Bd of Estimate of the City of New York,* 72 N.Y. 2d 674, at p. 682, 536 N.Y.S., 2d 33, 532 N.E. 2d 1261 (1988); under *H. O.M.E.S. v. NYS UDC,* 69 A.D. 2d 222, 418 N.Y.S. 2d 827 (4th Dep't 1979), the DEC must identify the relevant areas of environmental concern, take a hard look at them, and make a reasoned written elaboration of the DEC's decision about significance.

19. 6 N.Y. Code of Rules and Regulations, Section 617.7. See the checklist of issues for determining scope at 6 N.Y. Code of Rules and Regulations 617.21 (Appendix D).

20. Environmental Conservation Law, Section 8-0109(4).

21. Environmental Conservation Law, Section 8-0109(7)(a).

22. Environmental Conservation Law, Section 8-0109(7)(b), and vol. 6 N.Y. Code of Rules and Regulations Section 618.1.

23. Environmental Conservation Law Section 8-0103(2) and vol. 6 N.Y. Code of Rules and Regulations Section 617.14(a).

24. *Jackson v. NYS UDC,* 67 N.Y. 2d 400, at p. 415, 503 N.Y.S. 2d 298, 494 N.E. 2d 429 (1986).

25. Environmental Conservation Law Section 8-109, and vol. 6 N.Y. Codes of Rules and Regulations Section 617.14(i).

26. SEQRA requires that the findings cover each of the elements in Environmental Conservation Law 8-0109 and vol. 6 N.Y. Code of Rules and Regulations Section 617(9).

27. Environmental Conservation Law 8-0111(1) and (2).

28. See *John H. Alden et al. v. Maine,* U.S., 119 S. Ct. 2240, 67 U.S.L.W. 3683 (June 23, 1999), holding that unless a state agrees to be sued it cannot be sued for alleged violations of the federal Fair Labor Standards Act.

29. See vol. 40 Code of Federal Regulations Part 1500.

30. Environmental Conservation Law, Section 8-0105(6).

31. L. 1973, ch. 790; Environmental Conservation Law Article 25.

32. Section 1 of L. 1973, ch. 790.

33. Environmental Conservation Law, Section 8-0101.

Chapter Ten

Legal and Policy Challenges of Environmental Restoration

Joseph L. Sax

How does one begin thinking about an issue such as wolf reintroduction and its place in our legal system? The most obvious legal issue that comes to mind is that introducing wolves into any sort of settled area creates a risk to property, primarily to livestock and pets, and potentially to other interests, such as hunting. If one examines the usual public debate in communities where restoration programs are proposed, it is notable that opposition rarely is eliminated by offers of compensation or by evidence that the level of risk is low. Why is this? There are several possibilities, of course. People may disbelieve statistics or may conclude that compensation programs are ineffective because the burden of proving wolf depredation is too difficult. I suggest that these are only the most conspicuous concerns and that they are undergirded by a much more far-reaching unease. Simply stated, environmental restoration has dramatic and disquieting implications for the legal system's agenda for land, as it has developed over centuries.

What has that agenda been? The conventional understanding is that private land is there to be used for economically productive purposes: agriculture, stock-raising, residences, and industrial and commercial (including recreational and leisure) purposes. To introduce programs that intrude on these purposes, by ordinary proprietary standards, is a diversion from land's "productive use" mission. The wolf may be only a small diversion, but in the conventional perspective it is symptomatic of something drastic and disconcerting. To make everyone accommodate the wolf, rather than the other way around, is to turn things upside down, to put in question the very purpose of the private property system in land. This is not to say that conventional thinking opposes wildlife and wilderness or environmental protection. Rather, ordinary thinking simply assumes that such programs are a function of the public lands and that the values associated with them should be provided in places such as parks and wildlife refuges.

Conventional thinking is comfortable with the division of function that generated our existing system of separate enclaves set aside for purposes such as refuges. What is threatening to owners today is the implication that such lines will be erased and that private owners will, to some extent, be called on to play what has been thought to be an exclusively public role. The fear is that private rights will be eroded away and that a fundamental distinction between private and public will be lost. Although the arbitrary lines that separate the private from the public realm obviously are irrelevant from the perspective of habitat, the blurring of that frontier as a legal and policy matter is precisely what alarms property owners and commodity users of land. All this is not to say that the anxieties I have just described are well founded, either in history or in prospect, but only that they exist and are powerfully felt. However, they do not arise from mere imaginings or wishful thinking.

The Nature of Property Rights

Under the basic precepts of our legal system, an owner of land bears no responsibility to maintain or support any of the natural benefits of that land.[1] In fact, owners may denature it as fully as they want. To take a familiar example, an owner may cut down an ancient redwood forest, and if the public wants to prevent such activity, it must buy that protection from the owner. At the time of this writing, the press was carrying stories about negotiations in which the owner of old-growth redwoods in the Headwaters in northern California was to get nearly $500 million to prevent him from logging that forest (*San Francisco Examiner,* Editorial 1998, A20). Not only do owners not bear affirmative obligations to maintain natural values, but during at least the past several hundred years, the law positively encouraged denaturing.

For more than a century, it was the general rule in western water law that the only way to obtain a property right in water was to abstract it from a river, for that alone was defined as a beneficial use (e.g., *Steptoe Live Stock Co. v. Gulley* 1931). No one could obtain a right to maintain instream flows, whether for recreation, preservation of fisheries, or protection of ecological values. As a result, many western streams were completely dewatered and left as dry beds, especially during the summer, when water was needed primarily for agricultural irrigation. In addition, it was impossible to acquire a right to water to maintain it instream, even where one wanted to do so to maintain a river as habitat for fish or for recreation. A leading case decided in 1913 is illustrative. The complainant owned a resort and wanted—as the court put it—to maintain "the continued natural falls and flow of the stream" for the benefit of its guests, for recreation and fishing. The court rejected this application:

> The laws of Colorado are designed to prevent waste of a most valuable but limited natural resource. . . . They deny the right of

the landowner to have the stream run in its natural way. . . . The state laws proceed upon more material lines. . . . The dominant idea [of the law] was utility, liberally and not narrowly regarded. (*Empire Water & Power Co. v. Cascade Town Co.* 1913)

That was the dilemma of anyone who wanted to maintain the natural values of a stream in the West for many decades. As recently as 1974, a statute in Idaho enacted to permit the state to set aside a stream in its natural condition and protect it from further diversions was attacked as being unconstitutional and sharply split the justices of the Idaho Supreme Court (*State Dept. of Parks v. Idaho Dept. of Water Administration* 1974). The rules of that old 1913 case generally no longer apply, and water can be set aside to maintain instream flows, but several points must be noted. First, in all the years that the old rules were in place, the bulk of the water in the West went into private ownership, and the users now have constitutionally protected property rights in that water. The new rules apply only to streams that haven't already been significantly denatured. Although the constitutional rights of existing users are not totally fixed and are subject to certain public claims, at the very least efforts to restore (rather than maintain what still exists of less disturbed systems) raise very difficult problems of modifying existing uses on which individuals and communities have long relied.

Similarly, under laws such as the General Swamp Land Act of 1850, wetlands were granted into state and then into private ownership for the express purpose of getting them filled and developed (that is, "beneficially used"), thus destroying them as wetlands (Gates 1968). Although some statutory inroads have been made on these property precepts, they are highly controversial and continue to be the subject of legal challenge, particularly in wetland protection laws such as the Clean Water Act and the Endangered Species Act (ESA) restrictions on the owners of listed species' habitat. But even more significantly, the old rules largely remain in effect. Outside the strictures of the ESA, owners still are generally free to modify their land so as to make it useless as habitat (e.g., by fencing, filling, or deforesting). That is precisely what has happened to the bulk of privately owned land in this country and a good deal of publicly owned land.

It is not an exaggeration to say that the history of land use in the United States over the past several hundred years could be written as a story of the denaturing of the landscapes and waterscapes of the nation: forests turned into cities, prairies converted to farmland, rivers dried up or contaminated as refuse conduits, and wildlife extirpated in favor of livestock. This is the tale told so elegantly about early New England by William Cronon in his book *Changes in the Land* (1983): the conquest (that is, the destruction) of the wilderness. I do not need to recount that history here; it is eminently familiar. What is perhaps not so familiar, or at least not so obvious, is that our property system has been in full accord with that history and with the goals that underlay it. That is, our system of property rights permitted, encouraged, and rewarded activities that denatured the land and imposed no restrictions or penalties on those who did so. Nor did the system create any obligations to maintain or facilitate natural systems.

All this does not mean that landowners had absolute freedom to do what they wanted. On the contrary, there were always many restrictions on land use, but those restrictions were designed almost exclusively to protect human interests. For example, the centuries-old law of nuisance restricted polluting activities that threatened human health or human amenities or that transgressed moral standards. There were positive obligations, such as rules against leaving land uncultivated when the community needed production, or protection of certain forest lands whose trees were needed for naval ships' masts (Hart 1996). There were even laws protecting wildlife, and they did sometimes protect animals for their own sake, but usually such laws were designed to maintain viable populations for the benefit of hunters (Field 1984; Lueck 1989).

What is far more difficult to find is any restriction protecting natural functions as such. There seems to have been no notion that the natural world provided important values (such as biological diversity), which were a kind of capital asset that was of value to everyone. Had those values been recognized, a kind of duality in land undoubtedly would have been acknowledged; that is, that although land was a resource that could be parceled out and used for the private benefit of the owner and for the collective benefits we all got from such use and development (wood to build houses, for example), it was also needed for its natural services. Had this duality been recognized, notions that land was capable of total privatization (so total, it was thought, as even to include the right to destroy)[2] would never have become prevalent. It would hardly have been thought that an individual would own the right to destroy biological diversity.

Of course, there is an older tradition of commons in which there was a sharing of the natural products of the land, such as the reeds, peat, and fish that medieval marshes supplied,[3] but that tradition (itself economic rather than preservationist in its purposes) died out after the enclosure movement in seventeenth-century England, and today the essential notion of property that people everywhere carry in their heads is the postcommons one of privatization and development I described earlier.

In any event, the oversight—if it can be called that—in not providing a legal concept to protect collective heritage was not limited to the natural world. It was perhaps even more glaringly obvious in the realm of cultural property.

Cultural Property

Although the idea of preservation seems to have first appeared in Western Europe around 1500 (Lynch 1972), the notion that something could be private property and at the same time be of such importance to the community at large that the private owner could be obliged to protect it for the benefit of the community—as a legal obligation—did not come into being until the nineteenth century. The setting for that novel idea was historic preservation, exemplified by Stonehenge. At that time Stonehenge was located on privately owned land, and it belonged—in the fullest sense of the word—to the landowner, who was perfectly free under English property law to do whatever he wanted with this invaluable object of historic and scientific interest.[4]

Not only could owners do what they wanted, but in the setting of a building boom, such monuments were being destroyed both in England and on the European continent. Although it seems incomprehensible today, some of the greatest structures in the world were being taken down for the value of their stones as building material. Perhaps someday the destruction of wildlife habitat by landowners will seem as bewilderingly inappropriate.

Although there had always been some public things, such as the sea, that were available to the whole community (Deveney 1976; Sax 1970), the notion of a conventional private thing (such as the land attached to an individual's country home) being treated as if there were some public claim on it was very unfamiliar. Indeed, the debate that occurred in the English Parliament when the first historic preservation law was proposed in the 1880s is of extraordinary interest, for it foreshadows the sort of rhetoric that accompanies controversies over identifying private land today as environmentally precious habitat. When Sir John Lubbock introduced his bill to protect ancient monuments just over a hundred years ago, he explained that the nation's greatest monuments "have generally been sacrificed, from ignorance of their value and interest, for the most trivial reasons. They have been carted away to manure the ground, or broken up to mend the roads. At present there is no one who has the right, in the name of the nation, to say a word to prevent such acts of vandalism."[5]

Lubbock's bill was terrifically controversial, not because the English did not care about history but because they could not conceive of a public right in what seemed ordinary private property. For example, it was reported in the House of Commons that the owner of Stonehenge "objected to having the jurisdiction over his own property taken out of his hands."[6] The reason, it was explained, was that under the proposed law "the proprietor was dealt with not as the owner of his property, but as a mere trustee of it. They might as well deal in that way with an old picture which had an owner as with an ancient monument which had come down to him with the family estate" (Kains-Jackson 1880, 109, quoting Earl de la Warr).

The attorney general similarly objected to the proposal, a quite limited law to protect only ancient monuments such as Stonehenge and Avebury. He said,

> If they adopted the principle of the Bill . . . where was its application to cease? If they were going to preserve at the expense of private rights everything which happened to be of interest to the public, why should they confine the legislation to those ancient monuments? Why should they not equally provide for preservation of the medieval monuments—of those old abbeys and castles which were quite as interesting as Druidical remains? And why should they stop even there? Why not impose restrictions on the owners of pictures or statues which might be of great national interest? . . . Might they not also say that a certain row of beech trees on a man's estate which gave great pleasure to persons passing by ought in the same manner to be preserved?[7]

Historic preservation ordinances are by now so familiar to us that these statements must seem as if they came out of some very remote and very primitive past. Yet the notion that the owners of things that have great value to the community owe some sort of affirmative obligation to accommodate the public interest in them is still an unfamiliar one in most realms, particularly but by no means exclusively the environmental domain. A standard text on art law contains the following passage: "An eccentric American collector who, for a Saturday evening's amusement, invited his friends to play darts using his Rembrandt portrait as the target would neither violate any public law nor be subject to any private restraint" (Feldman and Weil 1986, 434).[8] That is an accurate statement of contemporary law. The notion that we, collectively, have a right to protect even the greatest artistic masterwork in the world from an owner's caprice or neglect is even today at or beyond the borders of legal theory (Sax 1999).

The same is true for artifacts of scientific importance. You may recall the press coverage only a few years ago of the excavation of the dinosaur "Sue," described as the largest and most intact example of *Tyrannosaurus rex* ever found. Because the fossil was on private land, it belonged to the landowner in exactly the same way oil or iron under the surface soil is owned. The potential scientific importance of the object, its significance for research that could add to the total of human understanding, is of no legal significance. It was only good fortune that when the fossil was put up for auction at Sotheby's, two corporations, McDonald's and Walt Disney World Resort, put up the funds to purchase it for the collection of the Museum of Natural History in Chicago. Had they not done so, as *Time Magazine* pointed out at the time, "Bill Gates could buy her on a whim. So, for that matter, could Steven Spielberg, Michael Crichton or Madonna. She would make a terrific conversation piece. No one knows what will happen. . . . She may end up . . . as a lawn ornament of the rich and famous" (Lemonick 1997, 74).

These examples illustrate one of the great puzzlements of our law: the near absence of any notion of a collective endowment that the present is duty bound to protect and pass on to the future. Protection of our natural heritage (as contrasted with the abatement of nuisances) dates only from the first forest reserves established in the 1880s and 1890s and the Yellowstone National Park Act of 1872. Important and pathbreaking as they were, those laws never challenged the fully privatized conceptions of our property system in the way in which Sir John Lubbock's bill for protecting ancient monuments did. Our early nature protection schemes created no affirmative duties; they essentially set aside enclaves and reserves of one kind or another. One might even say that in setting aside certain areas, they legitimated—or at least eased the public conscience about—the continuing destruction of biological diversity everywhere else. Not until the late 1960s and early 1970s, a mere few decades ago, did notions such as maintaining wetlands[9] or protecting species[10] and the recognition that all our lands, public as well as private, were implicated in those tasks get onto the public policy agenda.

Despite extensive efforts, I have been able to find in our legal tradition almost no notion of a public duty to protect and pass on either our natural or cultural heritage unimpaired to the future. I have spent a good deal of time looking at the governance of human-made things (cultural property) because it has a longer history of appreciation than does nature preservation and because I thought it might reveal principles that could be applied to nature preservation. The dearth of authority has been surprising to me.

For example, one might have thought that the ancient law governing heirlooms, though a part of the private law world, would provide a concept of societal continuity, focused on family, giving rise to a duty of preservation, protection, and legacy through time. However, that is not the case. Although the law does govern rights of inheritance in heirlooms, there seems to be nothing to prevent the owner of an heirloom during his or her lifetime from doing with it whatever the owner wants, including destroying it or selling it and spending the proceeds. On a public scale, the same thing is true of public heirlooms, such as the crown jewels. Although these seem the very essence of a public treasure to be passed on from one to the next monarch, and although that has been the ordinary course, there are many instances in which such objects have been melted down, dismantled, or sold for their economic value, with the proceeds used to support military ventures of one kind or another. It seems to be assumed by legal authorities that such decisions were within the authority of the monarch and that there was no duty to pass along even the royal patrimony undiminished to the next generation, although I suppose that any disposition these days would at the least generate a terrific scandal.

The very idea of a public heritage is a modern notion in several respects. For one thing, a "public" in the modern sense—as we think of a public museum or an institution open to the public—dates only to the late eighteenth century. For example, the Louvre was a royal palace before it became a public museum. Our notion of preserving historic structures was unknown to earlier times. Churches were rebuilt, built upon, or built over as needed. The Gothic was not admired in the eighteenth century; our appreciation of it is largely the product of nineteenth-century sensibilities. Iconoclasm was a standard practice, both religious and political, whereby the monuments of other regimes and other beliefs were struck down as a dramatic form of disapproval. Far from being simply irrational outbursts, iconoclasm is the most logical behavior. Even today one could hardly imagine modern Germany with statues of Hitler, kept simply because they are artful. When the Abbé Grégoire spoke out in the 1790s against the destructiveness of the French Revolution and in favor of preserving the nation's artistic heritage—though revealed largely through work that was done under the patronage of, and for the glorification of, the ancient regime—he was expressing unprecedented and radical views (Sax 1990b, 1155):

> Certainly the temple of the Druids at Montmorillon, and that of Diana at Nimes, were not built by the hand of reason; and nevertheless is there any true friend of the arts who would not want them to be preserved in their entirety? Because the pyramids of Egypt had been built by tyranny and for tyranny, ought these monuments of antiquity to be demolished?

Although it is familiar now, the image of cultural artifacts as common intellectual and aesthetic assets was novel then. When Grégoire spoke of "the productions of genius and the means of instruction [as] common property," "national objects which, belonging to no one, are the property of all," he was talking in a way that had almost never been heard before, putting forward a conception of things that inherently belonged to the nation as a whole. To be sure, there were intimations of such ideas previously, perhaps most famously in Raphael's celebrated letter to Pope Leo X written around 1519 in which he lamented the ongoing destruction of ancient Roman relics, then being sacrificed to a sixteenth-century building boom (Sax 1990b). For the most part, however, it wasn't until the time of the French Revolution that terms such as *common property* and *common heritage,* which appear in his discourses, were heard. They were then picked up in the next century by writers such as Victor Hugo in France and John Ruskin in England, by which route they have become our common currency and the source of contemporary commitments to historic preservation.

It has often been pointed out, and certainly correctly, that ideas of preservation and heritage, and ultimately of what the present owes to the future, were products of a world that was changing, a world that became dramatically aware (perhaps for the first time) of large and permanent loss, of the capacity of people and their machines to bring about irrevocable change, so that the earth would never again look the same (Lowenthal 1985).

Paradoxically, however, the same forces that gave rise to concerns about saving a world that seemed to be succumbing to voracious industrialization also generated ideas of progress, conquest, and inevitability that threatened to sweep aside everything in their path. People during the last two centuries have lived largely by an ideology that required and even idealized a transformational economy that ensured that future generations would live in a world nothing like that which had gone before (Sax 1993). Forests were turned into fields for the plow; wildlife was exterminated in favor of domestic herds; rivers were dammed for hydropower, first for mills, then for irrigation and electricity and cities; prairies were extirpated; and wetlands were drained and filled to support productive agriculture. I say "productive agriculture" because those who brought about these changes believed—as John Locke had explained—that they were generating productivity in a void. After all, the displacement of Native Americans by European settlers was legally justified on the Lockean notion that the American continent was largely unoccupied, "empty space," a notion that completely misconceived the economy and ecology of the indigenous peoples who inhabited the New World—which is to say the productivity of a landscape that had not been subdued according to European notions of use and productivity. The very difference between what they saw (swamp and other "waste" lands) and what we see (biologically productive wetlands) explains why the notion of retaining biologically diverse habitat as heritage, as a productive capital good, as an inheritance we owe to the future lest we spend our children's endowment, was not a notion that our predecessors conceived.

This is not to say that no notion of heritage and no transmissive traditions existed. They did, in all sorts of ways. Schools passed on knowledge; literary traditions, written and oral, functioned; religious teaching was communicated through the generations; collectors have flourished for centuries; and a great richness of artifacts has survived over long periods of time. Yet with rare exceptions, legal systems have not been engaged with protection of heritage, natural or cultural. We are left with a dilemma and a challenge. We recognize a need to address these issues with the assistance of the legal system, but we have very little experience or tradition that we can look to for guidance or for some theory on which to build.

The U.S. Constitution speaks of a responsibility to the future in only one place. The preamble, in language familiar to every schoolchild, provides that the Constitution is established to form a more perfect Union, to establish justice, ensure domestic tranquility, provide for the common defense, promote the general welfare, and "secure the Blessings of Liberty to ourselves and our Posterity." That is the first and last time the Constitution acknowledges the claims of posterity. That, certainly, is not to suggest that the founders thought posterity should be left to take care of itself. It is simply to reiterate that our foundation document contains no notion of public heritage as an asset to be maintained and preserved and certainly gives no direction as to how contemporary issues such as environmental restoration should be addressed.

A profoundly important task lies before the legal community. We need to reconsider and redevelop some of our theoretical assumptions, and we need to begin building some practical, workable legal rules to implement the goals of biological capital maintenance and restoration. I would like to call attention to each of these tasks now, not exhaustively and not to resolve them, but rather to illustrate the nature of the task before us. First, and perhaps foremost, we need to do some serious and fundamental thinking about the nature of property and property rights.

Rethinking the Nature of Property and Property Rights

Perhaps the most difficult cases are those in which habitat has been significantly disturbed in pursuit of the goals of the transformational economy, and there is a felt need to undo some of what we have done. The Adirondacks appear to present one of the less troublesome situations because large parts of it remain undisturbed. Even where a good deal of development has occurred, however, restoration efforts can succeed—though hardly painlessly—as the recent resolution of the Mono Lake case in California shows (Ellis 1991). Far more problematic cases are those such as the Columbia River system, where under the traditional rules of use and diversion, a vast system of dams and reservoirs was constructed to promote irrigation and industry, particularly the aluminum industry, which uses large amounts of electricity and which benefited from inexpensive hydropower. The Snake and Columbia basin, once among the richest habitats for anadromous fish (salmon) is now the habitat of endangered species and the site of a huge complex of existing uses that affect navigation, electric power production, agriculture, municipal water supply, Indian fishing rights, ocean fisheries, and fish hatcheries. In the midst of all of this, we are confronted with such troublesome questions as whether barging fish around human-made barriers is the sort of "restoration" that ought to be accepted. Any level of restoration of the watershed as a viable habitat for salmon involves an extraordinary task of unraveling a century's creation of rights, relationships, and dependencies (Blumm et al. 1997).

Perplexing as the water situation may be, it nonetheless offers certain advantages over land in terms of shifting thinking about the nature of property rights. For one thing, water has always in some respects been considered a public resource rather than a purely private one. We say that one can acquire only usufructuary rights in water, rather than total ownership. It has always been recognized that certain public rights remained, such as the right of navigation, and to that extent public needs could trump purely private claims. These public servitudes help to smooth the transition from the purely transformational to a more mixed economy in the use of the nation's waters. In addition, although the facts vary from place to place, not all the waters have been taken up by private diverters and users, and the unappropriated waters that remain are in the public domain and subject to whatever new regimes or rules the public decides to apply (unlike land, which is all owned by somebody, whether or not it is being used for some economic purpose). Land lacks these theoretical limits. Except for tracts owned by the government in parks, forests, military establishments, and the like, the remainder is simply private property, with hardly any conception of ownership as somehow having to accommodate to habitat needs.

Perhaps the most revealing judicial opinion in explicating conventional thinking is that of Justice Antonin Scalia in the case of *Lucas v. South Carolina Coastal Council* (1992). As part of a shoreline protection program, the owner of an oceanfront tract was obliged to leave his land undeveloped. The case focused on the uncommon stipulated fact that the regulation completely devalued the land, but in the course of his opinion Justice Scalia reflected on a much more significant point, although it takes a bit of reading between the lines to get the full import of his concerns. He noted that most traditional regulation consisted of limitations on what an owner could do in the process of developing his or her land. For example, one might be limited in the density of development as to height or coverage of the land, a factory might be restricted in its emissions or in the noise it could make, a farmer might have to limit or change herbicides formerly applied. The common element in all these situations is that they assume economic use of the land by the owner and constrain that use where a product of it is in some respect harmful or unwanted.

On the other hand, the restriction in the *Lucas* case imposed an obligation of nonuse, not to control development but to permit the land to function in its natural condition. To the justice, this seemed entirely at odds with the very notion of ownership. How can you own land and not have a right to use it for *your* purposes (though with restraints and controls for the protection of others)? For, Justice Scalia said, quoting Blackstone, "what is land but the profits thereof?" (*Lucas v. South Carolina Coastal Council* 1992, 1017). Although the decision in the particular case turned on the loss of all value as a result of nonuse, I believe that a more fundamental issue troubled Justice Scalia and some of his colleagues: How can the public require an owner to devote any part of his or her property to nonuse simply to allow natural functions to prevail? To own land, as they see it, is to be free of the obligation to dedicate it as habitat.

I believe the reasoning would run something like this, although it is not at all explicit in the *Lucas* opinion. If one bought a 100-acre tract of land that was desirable and used as habitat for some animal because of the plants that grew on it in its natural state, and one wanted to plow up those plants and convert the tract to a farm, surely one could do that as part of one's ownership rights, even though the result would be to deprive the animals of that habitat. The reason is that "use," which is to say transformation from natural use to some human use, is exactly what ownership means. Although in such a hypothetical case the owner would reap no value if he or she had to keep the land in its natural state for the benefit of the animals, logically the owner is in the same situation, *pro tanto,* if only 10 of the 100 acres were the sort of habitat just described and if he or she plowed up that 10 acres and drove the animals off of it. Would not the owner's property rights be violated as to the 10 acres if he or she were required to leave them in their natural state for the benefit of the animals?

By the logic of the transformational economy, there is much force to the claim that such a restriction on even 10% of the land is an intrusion on property rights. The reason, of course, is that by the logic of that economy, natural services have no claim on the land and can be extirpated. The logic is precisely the same as that of the old water rights law. Because the only beneficial uses (property rights) are those that abstract water from the river, the fish who need the natural services of a river have no claim on any property owner, and thus, consistent with the logic and goals of the property system, they can be extirpated. Or, alternatively, if they are to be protected, the right of protection must be acquired back from property owners.

The logic is powerful only because it is the logic of a system that does not make protecting natural systems one of its goals. If it did—in the sense that it makes maintaining quiet enjoyment of residential lands one of its goals—it would define specified destruction of natural systems as a nuisance, just as it defines specified sorts of noise as a nuisance (i.e., as a limitation on the definition of property).

Can changes in the conception of property, as a result of changing goals, occur? Indeed they can. Examples of property law's adaptation to social changes abound. In a ruder world of the Middle Ages, nuisance law originally imposed unprecedented duties of neighborliness on owners' rights (Powell 1963). The Kentucky Constitution once opined that "the right of the owner of a slave to such slave, and its increase, is the same, and as inviolable as the right of the owner of any property whatever" (1850). In eighteenth-century America, the states abolished feudal tenures (Vance 1924), abrogated primogeniture and entails (Morris 1930), ended imprisonment for debt (*Sturges v.*

Crownshield 1819),[11] and significantly reduced rights of alienation (Powell 1963)[12] as well as dower and curtesy (*Ferry v. Spokane* 1922; *Randall v. Kreiger* 1874).[13] As the status of women changed, laws abolished husbands' property rights in their wives' estates (e.g., *Warburton v. White* 1899; *Baker's Ex'rs v. Kilgore* 1892; Donahue 1979).[14]

Earlier I used the example of historic preservation laws to show that protecting social capital in the form of cultural property was not a traditional element of our jurisprudence. As it became an important public goal during the last century or so, however, property rights have been reconceptualized to build preservation into the structure of property rights. None of this has happened very openly, and again it is necessary to read between the lines of important decisions to understand what is really happening. My example in this instance is the case of *Penn Central Trans. Co. v. City of New York,* a 1978 case in which for the first time the U.S. Supreme Court had to decide whether the owner of a historically or architecturally important structure bore an obligation to the public to preserve it, that is (analogously to the natural habitat case), to leave it as it is, even though that significantly restricts the owner's use and development opportunities and changes the presuppositions of the transformational economy.

Grand Central Terminal in New York was constructed in 1913. For more than half a century it had operated as a railroad station and as one of Manhattan's most familiar landmarks, situated at the intersection of 42nd Street and Park Avenue. In 1967, after a public hearing, it was designated under the city's Landmarks Preservation Law as an official landmark. The commission report described the station as "one of the great buildings of America. . . . It combines distinguished architecture with a brilliant engineering solution, wedded to one of the most fabulous railroad terminals of our time. Monumental in scale, this great building functions as well today as it did when built. In style, it represents the best of the French Beaux Arts" (*Penn Central Trans. Co. v. City of New York* 1978, 116).

In practical effect, Penn Central had to get permission from the commission to modify or demolish the station or to make any changes in its facade. In 1968, the Penn Central Company, to increase its income, entered into a contract to allow a 50-plus–story office tower to be built above the existing structure. Two alternative plans were designed by distinguished architect Marcel Breuer and submitted to the commission for approval. One called for the tower to be cantilevered atop the existing facade, resting on the roof of the terminal. The other would have required tearing down a portion of the terminal, including the 42nd Street facade facing Park Avenue. After 4 days of testimony and after hearing over 80 witnesses, the commission denied both proposals. "To balance a 55-story office tower above a flamboyant Beaux-Arts facade seems nothing more than an aesthetic joke," its decision said. "The 'addition' would be four times as high as the existing structure and would reduce the Landmark itself to the status of a curiosity" (*Penn Central Trans. Co. v. City of New York* 1978, 118).

Penn Central sued, claiming that the decision constituted a taking of its property without the payment of just compensation, in violation of its constitutional property rights. After fully pursuing its claims in the New York courts, Penn Central appealed to the U.S. Supreme Court. Surprising as it may seem, no landmark designation and property rights case had ever been heard by the Supreme Court. Although historic preservation ordinances had been common in the United States for many years—the Court noted that more than 500 of them were in force—most were of the sort that established historic districts, such as New Orleans Vieux Carré or Santa Fe's old town.

The essence of Penn's Central legal claim was twofold. First, traditional historic district ordinances were justified by a theory called mutuality of benefit and burden; that is, that although everyone in a district was restricted from modifying the buildings, each was also benefited by the restrictions imposed on his or her neighbors, which collectively attracted tourists and brought economic benefits to all. A similar theory underlies most zoning laws: Height or density is limited in an area, but everyone in the zone has to comply with the same restrictions to get the benefits of light and air or lessened congestion. As a result, the value of all property in the district is sustained. There was no such mutuality in the New York case, Penn Central said. The landmarking singled out an individual building such as Grand Central Station, and although its attractiveness may have benefited others, others bore no parallel burdens on its behalf. Indeed, owners of all the neighboring buildings were perfectly free to build skyscrapers and to reap the economic benefits, whereas Grand Central Station alone (but like other such designated structures in the city) had to remain as a low-rise landmark.

Penn Central's second argument was that the only justification for singling out a particular owner for restrictions was if it was doing something harmful to others. A particular polluting factory could be restricted, as could an unsafe construction site or a nightclub that was making noise and disturbing its neighbors. Penn Central was not engaged in any harmful activity in the sense of threatening safety, health, or morals. Penn Central emphasized that it had created no excessive density, noise, blocking of light, increased fire hazard, or the like. Indeed, it urged, all that distinguished it from other, nonregulated owners was that it had done something admirable.

The Supreme Court was not of a unanimous view. It divided 6 to 3. The majority of justices, joining an opinion written by Justice William Brennan, rejected Penn Central's claims and sustained the landmark designation ordinance. Justice Brennan and the majority of his colleagues viewed the case as a conventional one, citing both standard zoning and regulatory decisions but never really addressing Penn Central's point that all those cases rested on some notion of mutuality or on some claim of harm. None involved the mere withdrawal of a benefit that had previously been conferred, such as the demolition of an architecturally significant building.

Chief Justice Rehnquist wrote a dissent. His opinion evinced a strong sense that *Penn Central* was fundamentally unlike an ordinary zoning case, although he did not elaborate the point. However, he did make one striking and ironic observation: "Penn Central is prevented from further developing its property," he said, "basically because *too good* a job was done in designing and building it" (*Penn Central Trans. Co. v. City of New York* 1978, 146). The point is a telling one. In the ordinary case, obligation arises only because the owner has done something undesirable. Justice Rehnquist pointed out that the sole reason Penn Central was worse off than its neighbors was that it had designed and built an especially fine building and that it now wanted to withdraw the benefit that its presence had conferred on its neighbors. Notably, nothing in the proposed demolition itself nor in the increased density Penn Central proposed was prohibited. If the existing building had been an undistinguished one, Penn Central would have been perfectly free to demolish it and build a 50-story replacement without running afoul of the law.

The conclusion implicit in the Rehnquist opinion is simply that one who owns something deemed especially valuable to the community, by virtue of that fact and of his or her previous gift of the benefit of that boon to the neighboring community, has somehow incurred an obligation to protect or preserve the object for the benefit of the community. Justice Rehnquist called this special sort of obligation "an affirmative duty to preserve" (*Penn Central Trans. Co. v. City of New York* 1978, 141, 146), contrasting it with negative duties to refrain from doing harm. He also suggested that one of the ordinary elements of a property right, the right to "dispose" of one's own property, had been withheld from Penn Central.[15]

Rehnquist is correct in observing that the *Penn Central* case seems to change the conventional view of the rights and responsibilities of owners and to suggest a new sort of affirmative obligation. Moreover, the benefit Penn Central is required to maintain and confer is its own creation, which would not be there but for its commitment to quality and good taste. It is not the beneficiary of mere luck, finding oil beneath its land, or the possessor of a bounty of nature to which it has added nothing, such as a grove of ancient redwood trees. Nor is it claiming something like ancient fossils or the artifacts of an ancient civilization, to which it has contributed nothing. By ordinary standards, it is an unlikely candidate to be subjected to government restriction.

In light of this, one might have expected the Supreme Court to acknowledge that the *Penn Central* case was staking out previously unexplored territory. For whatever reasons, however, the majority did not take up Justice Rehnquist's challenge. Perhaps it was not ready to face the significance of conceding that the public has a stake in the fate of objects of "special historic, cultural, or architectural significance" (*Penn Central Trans. Co. v. City of New York* 1978, 108). The legal system apparently is not quite ready to confront such an idea in principle. So Justice Rehnquist's challenge goes unacknowledged and unanswered. The owner-as-custodian, owner-as-steward, remains the law's awkward little secret.

Conclusion: Legal Protections for Natural Capital

The sort of affirmative obligation toward our cultural capital that the Supreme Court, however covertly, recognized in the *Penn Central* case is what we will have to apply to protecting our natural capital if we are to move significantly in the direction of maintaining and restoring biological diversity and moving beyond the publicly owned enclave approach that has previously characterized American nature preservation policy. Fortunately, wildlife protection is one of the few areas in which the notion of a public entitlement has long had some currency. There are at least a handful of prominent court decisions around the country in which a state, to protect its wildlife populations, has imposed restrictions on hunting or otherwise interfering with animals. Predictably, these rules sometimes resulted in damage to property. Perhaps the most famous such case arose in the State of New York.

In 1900, in response to the near extermination of its beaver populations, the New York legislature enacted a law prohibiting the hunting, molestation, or disturbance of beavers. A few years later it acquired a number of beavers and began restocking certain Adirondack streams with them. One of those streams, where the new population flourished, happened to abut a tract of land held by a Mr. Barrett. The land was valuable for building sites because it was amply forested. The planted beavers assiduously felled hundreds of Barrett's trees, reducing the value of his land. The Barretts were barred by law from molesting the beavers. Because the state had introduced the beavers, Barrett sued for the damage the trespassing animals—effectively as agents of the state, according to his lawsuit—had done. The highest New York court rejected his demand for compensation. Surely, the court said, the government has a right to protect wild animals, a right that had been recognized in colonial laws going back to the early 1700s. So, said the court, there was no doubt of the validity of the hunting and harming ban. And, it continued, "wherever protection is accorded, harm may be done to the individual. Deer or moose may browse on his crops, mink or skunks kill his chickens; robins eat his cherries . . . and no one can complain of [these] incidental injuries that may result" because they are the inevitable consequence of the valid policy of wildlife protection. It then went on to make a remarkable statement:

> The public authority is not to be limited to guarding merely the physical or material interests of the citizen. . . . The eagle is preserved not for its use, but for its beauty. . . . The same thing may be said of the beaver. . . . Observation of the animals at work or play is a source of never-failing interest and instruction. . . . If they cause more damage than deer or moose, the degree of the mischief done by them is not so much greater or so different as to require . . . a special rule. . . . [Their] preservation does not unduly oppress individuals. . . . The prohibition against disturbing [the beaver] is no . . . different . . . from that assumed by the Legislature when it prohibits the destruction of the nests and eggs of wild birds even when the latter are found upon private property. (*Barrett v. State* 1917)

This is truly extraordinary language and it has influenced courts in a number of cases, including some modern ones. A 1968 Montana case, in which protected elk were coming on an owner's land and eating his pasture, made a very important observation. After noting that Montana was a state where wild game abounds and is regarded as one of the state's greatest resources, the court said that "one who acquires property in Montana does so with notice and knowledge of the presence of wild game and presumably is cognizant of its natural habitat. . . . Accordingly a property owner in this state must recognize the fact that there may be some injury to property or inconvenience from wild game for which there is no recourse" (*State ex rel. Sackman v. State Fish & Game Comm'n* 1968, 666). There are now several excellent studies collecting similar authorities around the country.[16]

These cases are very encouraging precedents for any effort to implement a plan of restoration, and they benefit from both the traditional notion that wildlife is owned by the state for the benefit of the public and from the early recognition that scarcity necessitated regulation (as in the form of hunting seasons and bag limits; habitat protection came later). The cases also make clear that the protections involved were more "inconveniences," as the *Barrett* case put it, than deep incursions on economic uses of land. In light of what has been said in this chapter about pervasive attitudes about land and proprietorship, it is important to recognize that restoration efforts—despite the presence of some favorable precedent—are likely to encounter substantial resistance reflecting deep-seated attitudes and values about land and what it is for. The more restoration efforts appear to threaten established and economically important interests, the greater the challenge of finding new accommodations between the transformational, industrial, commercial economy and the economy of nature.

The magnitude of the task cannot be overstated. As I noted in referring to the salmon issues in the Snake and Columbia basin, restoration ordinarily presents more difficult problems than maintaining existing conditions, simply because it involves disrupting existing patterns of use and developmental expectations and calls for greater and at times more disturbing adaptations.

Whether the task is restoration or maintenance, however, it cuts deeply into long-standing legal conceptions that are deeply at odds with it. We cannot continue on the path of the old transformational economy, nor can we return to an economy of nature that reflects a hunting and gathering society or any other form of nonindustrialized (or industrialized) world. The question is, What sort of adaptations will permit land and water to serve agricultural, industrial, recreational, and municipal needs while sustaining a more desired level of biological diversity than result from the paths we have been on? I am no expert on these matters and would not assume to describe any formula or formulation. However, I see some encouraging steps that may point us in a positive and hopeful direction. The work done on habitat conservation plans under the aegis of the federal ESA is designed to find accommodations that permit economically productive work to go forward—in developing urban areas as well as in forests suitable for commercial timber harvesting and agricultural lands—while maintaining and restoring species to protect biodiversity (Turner and Rylander 1997). Some of the strategies seem straightforward and can be envisioned as elements of an emerging new notion of land and water rights in which both transformational and natural benefits can be reaped. Among them are reducing overgrazing, withdrawing from riparian areas in timber harvesting, clustering development away from sensitive areas, banning road building on steep slopes, restoring disturbed wetlands, encouraging market transfers of agricultural water with commitment to some marketed water to instream flow restoration, allowing less sprawling development into existing habitat, and the like. From what I have seen so far, the task is likely to be more one of moderation, adaptation, and accommodation than of uprooting, chaos, and upheaval.

Endnotes

1. One finds occasional exceptions (e.g., *Just v. Marinette County* 1972; *Babbitt v. Sweet Home* 1995; *Palila v. Hawaii Dept. of Land and Nat. Res.* 1979, 1981).
2. It is uncertain whether there has ever been a right to destroy. The *jus abutendi* (right of abuse) reflects the highly individualistic "will" theory of property (see Kersley 1949, 21–22). Writers on the civil law take a more measured view (Toullier 1824–1837, vol. III, p. 57).

3. An excellent recounting of the history is Bosselman (1996). See also Sax (1980).

4. The history is traced in Sax (1990a, 1990b).

5. 266 *Parlimentary Debates* (3d ser.) 885 (1882).

6. 237 *Parlimentary Debates* (3d ser.) 1983–84 (1878), p. 1549.

7. 232 *Parliamentary Debates* (3d ser.) 1542–43 (1877), p. 1551.

8. For a proposal to change the law, see Comment (1981).

9. Sec. 404(c), Clean Water Act, 42 U.S.C. sec. 1251, et seq.

10. Endangered Species Act, 16 U.S.C. sec. 1531, et seq.

11. Justifying abolition as not violating property rights.

12. Citing the rule against perpetuities, unlawful restraints on alienation, bars against antisocial dispositions, and insistence on formalities in dispositions.

13. Sustaining the abolition of dower against a takings claim.

14. *Warburton v. White* (1899) and *Baker's Ex'rs v. Kilgore* (1892), rejecting the claim that married women's property acts took the property of husbands. Under common law the wife's lands were subject to the husband's right to take the rents and profits during the marriage and to sell his interest in his wife's land without her consent (see also Donahue 1979).

15. *Penn Central Trans. Co. v. City of New York* 1978, p. 143. The dissenting judge in the Appellate Division said that imposing a duty to maintain the station "is to clearly lessen Penn Central's estate. . . . Otherwise stated, ownership entitles one to destroy as well as to preserve" (*Penn Central Trans. Co. v. City of New York*, 1975, Lupiano, J., dissenting).

16. Among especially interesting cases are *Christy v. Hodel* 1988 (grizzly bears), and *Southview Associates v. Bongartz* 1992 (residential development denied in deer yard). Also see Houck (1976), Babcock (1998, citing many authorities), and Sax (1998).

Part IV

Lessons Learned and Applied

Wolf reintroduction in Yellowstone National Park and reintroductions and natural recovery in the Great Lakes states provide an important backdrop to the Adirondack proposal. Thus, to situate the present proposal in terms of wolf biology and the politics and sociology of the deliberative process, the three chapters in this part focus on lessons that have been learned and that might be applied to the Adirondacks.

Chapter Eleven, "Lessons from Wolf Restoration in Greater Yellowstone," is by Tim Clark and Anne-Marie Gillesberg. Clark is professor (adjunct) in the School of Forestry and Environmental Studies at Yale University and board president of the Northern Rockies Conservation Cooperative in Jackson, Wyoming. Gillesberg does field work at the Northern Rockies Conservation Cooperative.

Drawing on the Yellowstone experience, Clark and Gillesberg examine the notion of common interest in governance of democratic society, and what is meant by learning or finding lessons.

According to these authors, common interests evolve from a democratic process that is open, reliable, fair, and honest and reflects the interests of a majority of Americans. Because a variety of special interests are involved in wolf restoration questions, the common interest is difficult to identify and sustain. Using the six phases of a reintroduction decision process—initiating recovery efforts, examining the problem and possible solutions, determining program direction, implementing the program, evaluating activities, and terminating or redirecting the program—Clark and Gillesberg offer an assessment of the factors necessary to achieving a technically and democratically informed deliberative process. They identify the following lessons. First, endangered species restoration in general, and wolf reintroduction in particular, is an involved process requiring the long-term commitment of individuals and organizations and substantial monetary resources. The process can become contentious and highly politicized and is therefore not for the weak of heart.

Second, ultimately, social impacts will exert substantial influence on the success of the recovery program. Therefore, examining the social context within which recovery efforts will be embedded is recommended. Third, more attention to social context, specifically recruiting respected local and regional elites, may be necessary to encourage endorsement of the prescription. Fourth, technical considerations certainly are important, but emphasis on the education and involvement of sympathetic members of the opposition, especially at local and state levels, can limit hostility and reduce potential conflicts. Without local acceptance, however grudging, recovery can be compromised. Fifth, evaluating program success should not be limited to biological aspects but should include an ongoing assessment of the social ramifications. Increasing numbers of wolves may be a biological indicator of success, but strong trends toward renewed opposition to recovery may have much greater consequences for the long-term success of restoration efforts. Finally, changes in program direction or management can be difficult because participants often are emotionally invested in their roles in its development and committed to pursuing a particular course of action. Sensitivity to such difficulties is important.

All of these lessons point to the importance of a contextual, problem-solving approach to species reintroduction that promotes cooperative inquiry within an adversarial context and encourages participants to engage in reflective conversation about their values and attitudes. This systemic approach, the authors argue, offers the greatest opportunity to improve the process and achieve the common interest.

The next chapter in this part is by Rolf Peterson, professor of wildlife ecology at Michigan Technological University and team leader for the Gray Wolf (Eastern Population) Recovery Team appointed by the U.S. Fish and Wildlife Service.

In Chapter Twelve, "Wolves as Top Carnivores: New Faces in New Places," Peterson describes the wolf in its ecological and evolutionary contexts, using examples drawn especially from his long-term studies of wolves in Isle Royale National Park and recent data on the return of wolves to Michigan's Upper Peninsula. Wolf restoration is not simply the return of the wolf, he argues, but also the return of everything that wolves do, directly and indirectly, in an ecosystem. This includes the effects of wolves on herbivores and thus on vegetation, the effects on other carnivores in the competition for food, and the coevolution of wolves and prey.

Citing early work by Olson, Murie, Hairston, and others on food chain theory and the role of the wolf in limiting herbivores, Peterson observes that the direct influence of wolf predation on herbivores also will be manifested in indirect effects that ripple through entire ecosystems, potentially affecting growth and distribution of vegetation. That wolves affect tree growth rates, for example, by reducing moose density has recently been demonstrated at Isle

Royale. For a given area, the critical question is whether wolf predation significantly affects herbivore density and distribution. Wolf influence can range from minimal to substantial, depending on prey density, density of wolves and other predators, productivity of the habitat, and winter severity. In human-dominated landscapes, wolf effects can be reduced by food subsidies for herbivores (access to agriculture) and avoidance behavior by wolves.

An important implication of the food chain hypothesis is that top carnivores are regulated by competition for food. Thus we should expect to see evidence that the largest and most powerful carnivores might attempt to displace or otherwise get rid of smaller carnivores that expropriate food. This includes coyotes and dogs. The propensity of wolves to kill dogs is not an act of predation (direct food acquisition), says Peterson, but rather of competition. The evidence of this is that dog carcasses often are left uneaten by wolves. Such behavior does little to endear wolves to rural residents, but this evolved behavior pattern comes along with the real-world package we call the gray wolf, as does the historical tendency of wolves to kill domestic stock as prey.

One argument in favor of wolf restoration is that we are perpetuating the forces of natural selection that led to the development of big-game species as we know them today. Although long-term maintenance of species characteristics probably depends to a degree on continuation of similar selective forces, it may not be widely appreciated how prey behavior represents adaptations to living with wolves. The winter yarding behavior of deer, survival patterns of moose in relation to age, and fleetness of foot and precocial young all evolved as antipredator defenses.

The modern hunter, properly managed, can do a remarkably efficient and effective job of curtailing population increases and stabilizing cervid populations to achieve management objectives that are compatible with human societies. Nevertheless, Peterson observes, where it is possible to retain large carnivores together with prey, would it not be prudent to do so, in accord with the Leopold dictate to "save all the pieces" while trying to maintain wildland resources?

In Chapter Thirteen, "Reintroduction: Inspired Policy or Poor Conservation?" Christine Schadler, a Ph.D. candidate in natural resources and environmental history at the University of New Hampshire, draws on lessons from wolves in the Upper Peninsula of Michigan and panther restoration in south Florida to argue for a more holistic approach to habitat preservation than is presumed by the Endangered Species Act.

The endangered species model, she says, is a narrow and potentially counterproductive approach to broad problems involving the structure of institutions, economies, and human population. A holistic approach to the relationship between humans and the environment must focus on urban sprawl, sustainable agricultural practices, and human population control.

Citing William Cronon, Schadler observes that the prevailing view of humans as separate from "wild" nature is reinforced by the reintroduction of wolves back into parks and refuges. Such efforts communicate an idea about nature that sets us apart and thereby sets wolves apart from us. Natural recovery, on the other hand, has as its goal living *with* wolves under circumstances that are sustainable for long-term wolf survival. One of the important lessons learned from the wolf recolonization in the Great Lakes states, she says, is that its success can be attributed to the fact that the wolf came back on its own four feet.

With regard to the Adirondacks, Schadler acknowledges the significant geographic barriers to wolf immigration from Canada. Given this, there are two important alternatives to reintroducing wolves to the Adirondacks. The first is developing binational agreements to purchase easements on land between Canada and the United States to provide movement routes for many creatures. The second is to rethink the role that the coyote now plays in the Adirondack ecosystem.

According to Schadler, recent genetic studies in Canada by Theberge suggest that the wolf in Algonquin Park *(Canis lupus lycaon)* may in fact be the ancestral wolf of the Northeast. Even more surprising is the possibility that this wolf may be a red wolf. Depending on how these studies hold up, the small Algonquin wolf, *Canis lupus lycaon,* which averages only 55 to 90 pounds, may be the most likely candidate to be translocated to the Adirondacks.

Significant in this scenario is the fact that any reintroduced wolves would have to share an ecosystem currently saturated by the eastern coyote, *Canis latrans, var.* This coyote is larger than its western counterpart, averaging 40 to 60 pounds, with some individuals weighing more than 70 pounds. The eastern coyote may, in fact, be the result of interbreeding with the Algonquin wolf. This may have occurred when the coyote migrated east after the extermination of the wolf in New England at the turn of this century. This would explain their larger size, stronger jaws, and tendency to pack and to hunt deer effectively. These traits also may be the result of a rapid evolution taking place in the coyote population in the Northeast, which is selecting for larger size and greater sociality than those of its western counterpart, enabling it to take larger prey.

If the ancestral north woods canid turns out to be *Canis lupus lycaon,* returning this small wolf into a landscape fully occupied by the large eastern coyote raises serious biological questions. For example, if coyotes in the Northeast rely on deer as an important component of their diet, what will be the cumulative impact on the deer herd of two sympatric competitors sharing the same resource? Also, because these two predators are so close morphologically, what will prevent interspecific breeding between the two populations or genetic swamping of the small red wolf by the eastern coyote?

Perhaps the most important point, according to Schadler, is that the eastern coyote appears to be evolving rapidly to fill the niche of the wolf. We must remind ourselves, she says, that the eastern coyote hunts the fields and woods of the Northeast largely because of our actions 100 years ago. We cleared the land of forests, wolves, and their prey, moose and elk. With mostly cleared land, the deer returned and their population soared. With wolves removed and with little competition for this resource besides the human hunter, a new, opportunistic predator, the coyote, migrated in to fill the void we created. The coyote has now evolved in size and behavior into an increasingly more wolflike coyote. Perhaps the appropriate human response to the evolution of this predator, she says, would be to allow it to continue filling the niche that we now propose wolves to occupy.

Chapter Eleven

Lessons from Wolf Restoration in Greater Yellowstone

Timothy Clark and Anne-Marie Gillesberg

The reintroduction of wolves to Yellowstone National Park has been hailed as the most successful recovery effort in the history of the Endangered Species Act (ESA). Begun more than 20 years ago in an effort to restore the full complement of species to the nation's first national park, eventually realized through a commitment to cooperation and compromise, wolf recovery has captured the imagination of the American people and has enjoyed tremendous popularity. However, not all Americans are thrilled with the prospect of a recovered wolf population. Ranchers and other local interests, particularly in states such as Wyoming, view the return of wolves as a threat to their livelihoods and akin to a declaration of war on the West. Wolf recovery remains an emotionally charged management issue as the restoration process continues to unfold.

Finding practical lessons from this process to improve future wolf conservation in the Yellowstone area, the Adirondacks, and elsewhere is essential. There are many lessons from this wolf reintroduction effort, which continues to have a very high profile. This chapter briefly describes wolves in biology and society, governance in the common interest, and learning of lessons; offers and illustrates some practical lessons about the restoration process; and makes recommendations for current and proposed restoration efforts for wolves and other endangered species.

We have followed wolf recovery in Yellowstone closely for more than 20 years and offer our views on what has (or has not) been learned to date. We are particularly interested in how communities address questions, such as wolf reintroduction, that have important environmental consequences for the present and future. We are also keenly interested in democracy and current problems in governance that are especially evident in states around Yellowstone National Park but also apparent among New York State communities both within and outside Adirondack Park. Finally, we are committed to truly interdisciplinary approaches to problems of governance and natural resources management (Clark et al. 1996b; Clark 1997a, 1997b, 1999, 2000). Application of an interdisciplinary approach to the issue of wolf recovery necessarily entails a holistic examination of the full array of biological and social factors that influence the process and places these factors in the appropriate historical and cultural context. Armed with this integrated information and recognition of the particular importance of human values and perspectives, such problems are more likely to be identified, evaluated, and solved efficiently and effectively.

Wolves, Governance, and Learning: Finding Lessons

Lessons from the Yellowstone wolf restoration effort ultimately rest on the nature of the wolf's biology and society's perceptions, the notion of common interest in governance of democratic society, and what is meant by learning. This section briefly describes this perspective and provides a short history of the wolf restoration program in the Greater Yellowstone Ecosystem (GYE; Clark and Minta 1994; Clark 1999).

Finding Lessons from Wolf Restoration in Greater Yellowstone

Practical lessons have at their root a mixture of wolf biology and human cultural perceptions. As our knowledge and understanding of wolf biology have evolved, so have our perceptions. However, some beliefs are deeply ingrained and drawn from a steadfast adherence to accepted cultural myths. It is important to recognize how these myths have, in some cases, fed our fears and contributed to the public debate about wolf restoration and our responsibilities to the natural world as well as to future generations (see also Lee 1993; Gunderson et al. 1995).

THE WOLF IN BIOLOGY AND SOCIETY

The wolf is unique, both biologically and culturally. Biologically, wolves are fascinating animals displaying many adaptive features in reproduction, ecology, and social organization. Many aspects of wolf biology are described in Varley and Brewster (1992), the Northern Rocky Mountain Wolf Recovery Plan (USFWS 1987), the Final Environmental Impact Statement (USFWS 1994) concerning the reintroduction of wolves to Yellowstone National Park and central Idaho, and numerous other sources (e.g., Clark 1999).

Culturally, wolves hold a special place among animals. They haunt our imagination and are burdened with the images, dreams, fears, and stories that have accumulated over centuries (Casey and Clark 1996). They are the subjects of fables, hunting stories, folklore, and myths from around the world, as well as literary and psychological analyses. Indeed, enough words have been written about the wolf to fill a library, with the images and meanings represented in these writings too numerous to count (Casey and Clark 1996). Not surprisingly, the public debate about what the wolf is and what it represents takes many forms; including whether we, as a society, value the wolf as a species to be protected and restored to portions of its former range. At the heart of the debate is the old philosophical question about our relationship and responsibility to the natural world.

GOVERNANCE AND COMMON GROUND IN WOLF RECOVERY: AN ENDLESS PROCESS

Since its passage in 1973, the ESA has been reaffirmed by the American people several times. However, it is still not supported by some segments of society. This public interest policy has been and continues to be debated in theory and negotiated in practice. Similarly, finding the common interest or common ground in Yellowstone wolf recovery has been contentious and remains so (Klein 1998). This process in the GYE is being carried out in the field, in newspapers, and in the courts on a nearly daily basis. Wolf recovery, like all efforts to find the common interest, becomes a process of balancing and accommodating the diversity of culture, class, and personality in society (McDougal et al. 1981). The nature of the process and its outcome ultimately are determined by the diverse perspectives of the people who are involved, the values, resources, and strategies they bring to the process, and the specific outcomes they seek (Reading et al. 1996). People seeking lessons from this process, such as Adirondackers, should be guided by the idea of clarifying and securing the common interest in democracy. Outside observers may be able to help in this regard, and it is in this spirit that lessons learned from the ongoing wolf restoration effort in the GYE are offered.

We understand the common interest to evolve from a democratic process that is open, reliable, fair, and honest and reflects the interests of the majority of Americans. This process is a uniquely human endeavor through which participants share their particular perspectives and attempt to identify areas of overlap and arrive at some consensus. The ESA is one such process that represents a "noble human concern for other species," but it is vulnerable to less noble human traits such as aggressiveness and dogmatism, as well as domination by special interests (Clark and Brunner 1996, 1). Special interests are at play in the wolf restoration process under way in the GYE, to be sure. These interests have become apparent in preliminary consideration of the question of reintroduction to the Adirondacks and may or may not correspond with the common interest as it is revealed and determined over time.

Diverse interests may be principled (based on ethics), expedient (based on compromise), assumed (presumed to be in the common interest), or scientifically valid (supported by evidence) (McDougal et al. 1981) and are well documented in the contemporary written wolf record. The 51 stories retold by Casey and Clark (1996) also illustrate these interests and help ground the current debate about wolves in its historical and cultural context, moving it away from the vehement ranting of recent years towards a more informed, democratic discourse.

Learning from the Wolf Restoration Process

Lessons applicable to current and proposed recovery efforts are rooted in the concept of learning, particularly individual and organizational learning (Clark et al. 1994). Learning through the process of social inquiry begins with the introduction of new things and the new language used to describe them. Although new behavior may be the result of this exposure, the same basic rules, assumptions, and values may continue to operate as before. It is only when old assumptions begin to fall away and make room for new codes of conduct that new rules and accompanying behaviors can become established (Clark 1996a, 1996b).

Inquiry begins with the identification of a problem and a search for solutions. To be efficient and effective in this regard, there must be willingness on the part of the inquirer to step into the problematic situation and continually respond as the problem, and his or her understanding of it, evolves. The process of social inquiry is an interactive one that shapes the outcomes of actions and relies on the particular experience, intuitive knowledge and understanding, and awareness of its participants (Miller 1999). It focuses on the present context and, by definition, is adaptable. It is not bounded by traditional knowledge systems but is the outgrowth of interaction and therefore unique to a given situation (Schön 1983).

Because individual professionals are at the heart of the recovery process, individual learning is the key to addressing problems that arise. By questioning familiar assumptions and biases, individuals can respond and adapt to new situations, creating solutions that go beyond traditional and purely technical ways of doing things. But individuals generally work through organizations to accomplish recovery goals, and the existing network of institutional systems of authority and control typically rewards technical rationality and expertise, not reflection and learning. Thus it is not uncommon for professionals to find themselves confined to certain channels of operation, for the freedom to reflect, invent, and differentiate would threaten the order and constancy on which organizational life largely depends. For endangered species recovery efforts to be successful, then, organizations should be structured in such a way as to encourage and support open inquiry and foster what is known as double-loop learning. Such learning examines and addresses conflicts that challenge the operating assumptions and cultural norms of organizations and perhaps holds the most promise for improving prospects for wolf recovery in both the Adirondacks and the GYE (Schön 1983; Clark 1996a).

Finding the Common Interest in Wolf Restoration

The 1997 U.S. District Court decision to remove all reintroduced wolves and their offspring from Yellowstone National Park and central Idaho (*Wyoming Farm Bureau v. Babbitt* 1997) draws attention to the lack of clarification of the common interest in the process and the divisive debate that has ensued.[1] Although many participants are involved in the process of wolf recovery, environmentalists who support the return of the wolves to the GYE and members and representatives of the livestock industry who adamantly oppose reestablishment of the predator in their midst are featured prominently. Their perspectives differ markedly, but the situations in which they operate are surprisingly similar. Environmentalists maintain that wolves are a necessary and integral component of the GYE and are actively engaged in promoting wolf recovery through specific campaigns, political lobbying, media attention, and participation in public hearings. In comparison, agriculturists view wolves as a significant threat to livestock and, perhaps more importantly, a cherished way of life in the West. They operate primarily in the political arena by exerting their influence on conservative politics. Like environmental advocates, the livestock industry understands the power of the media and urges its constituency to capitalize on opportunities to express their opinions in public forums, effectively keeping open the debate and the question of whether wolves will remain in the GYE.

Wolf supporters subscribe to a moral standard that considers the goal of recovery to be unassailable. Organizations at the forefront of the debate are leaders in the conservation arena. They project images of confidence and power through their commitment to advocacy, activism, and the acquisition of knowledge needed to resolve recovery problems. In comparison, livestock producers seek to maintain their position of power and privilege in the conservative West and ward off threats to their economic status.

Those in favor of recovery use strategies of education, litigation, research, and advocacy to promote their position, whereas those opposed rely primarily on advocacy and litigation. In general, however, both groups seek to persuade decision-makers and the public through political channels. In an effort to move the process of wolf recovery forward, Defenders of Wildlife (a leader in predator protection) responded to the concerns of the agriculturists with an innovative program to compensate ranchers and farmers for losses of livestock to depredation by wolves. Such economic incentives have obvious appeal. However, although this program has been responsive to all incidents of confirmed depredation, ranchers claim that it cannot adequately compensate for indirect effects of harassment, such as reduced survivorship and fecundity. This argument is becoming increasingly familiar to supporters of reintroduction as the population of wolves in the GYE nears recovery.

Strategies used by the livestock industry have allowed its members to maintain their political power in the West. Similarly, those used by environmentalists and their organizations have resulted in the desired outcomes of increased awareness and support of the role of predators in maintaining viable ecosystems and the effects of a broader, more powerful public constituency. However, it is apparent from recent events that research results and educational programs, as well as innovative strategies used in an effort to secure the common interest, have not succeeded in alleviating ingrained fear and loathing of wolves. In addition, cooperation and compromise have pitted environmental groups against one another, with some preferring strict protectionist policies to programs that allow greater management flexibility.

Practical Lessons from Wolf Restoration in Greater Yellowstone

Restoring wolves in a way that supports the common interest requires carrying out a successful program. Such a program must be established, administered, and directed in the field as events demand. One useful way to conceive of a wolf restoration program, or any endangered species recovery program for that matter, is as a life cycle in which different phases or activities represent different developmental stages in the life of the program (Clark 1997b). These phases may include initiating recovery efforts, examining the problem and possible solutions, determining program direction, implementing the program, evaluating activities, and terminating or redirecting the program. Decisions in all six phases are made through attempts to balance legal requirements, technical considerations, and the need for political consensus. Close inspection of these decisions can reveal hidden dynamics of wolf recovery and provide valuable insight into the nature of the process by identifying where it needs improvement and determining potential points of intervention.

Initial appearances aside, program development does not necessarily proceed in a linear fashion. Several phases of the program can be under way at the same time, and feedback loops can result in revisiting particular stages of the life cycle. It comes as no surprise, then, that more than 20 years have passed since the Rocky Mountain Wolf Recovery Team was formed to address recovery. We offer lessons learned from each of the six phases based on our understanding of the Yellowstone wolf restoration effort, particularly the events of the last 3 years.

Lesson One

Lesson one stems from the initiation phase, in which people are focused on the problem of restoring wolves. During this preliminary stage, the problem is perceived, identified, and placed on the public agenda. Planning to achieve recovery goals also is begun. What program participants often fail to recognize, however, is that endangered species restoration in general, and wolf reintroduction in particular, is an involved process requiring the long-term commitment of individuals and organizations and substantial monetary resources. The process can become contentious and highly politicized and is not for the weak of heart.

It took years for the idea of restoring wolves to the GYE to evolve to the point of serious consideration. Aldo Leopold was the first to broach the subject in the early 1940s. Canadian wolf biologist Douglas Pimlott added his support for recovery in 1967 and also proposed the return of wolves to Banff and Jasper National Parks. Defenders of Wildlife was a vocal proponent of recovery at this time. But it was not until the creation of the ESA in 1973 and the subsequent listing of the Rocky Mountain gray wolf requiring the development of a recovery plan that these early advocates were joined by many others and the idea gained momentum.

The idea of wolf reintroduction grew out of a developing awareness of the role predators play in natural ecosystems. However, this role was not greatly appreciated by those outside of ecological professions. Predators were viewed largely with a mixture of fear and hatred by western conservatives, particularly members of the livestock industry whose livelihoods, they believed, depended on wolf eradication. They maintained that "predator control was necessary" and that the West was "doing fine without wolves" (Huffman 1993). Returning wolves to the GYE thus was met with staunch opposition from politically powerful foes, and the process faced great difficulty moving forward. For example, repeated calls for more research into the question of the feasibility of wolf reintroduction effectively stalled the process in the initial stages of development.

Lesson Two

Lesson two concerns examination of the problem of wolf reintroduction and identification of possible solutions. During this phase, the problem is defined in more detail using expert analysis and knowledge of technical considerations. Scientific investigations, such as feasibility studies, identify plausible options for responding to the problem based on its likely impacts and outcomes. A programmatic response is outlined and critical parameters are listed. Outcomes of this stage in the life cycle include gathering, processing, and disseminating all information relevant to the problem of wolf recovery. Alternative policies are identified and considered, and open, public debate is encouraged.

There is a distinct tendency to focus on the biological and technical aspects of wolf recovery (of which there are many) at the expense of a full consideration of the social impacts and an integrated look at both dimensions. Ultimately, social impacts influence the success of the recovery program. Therefore, the social context within which recovery efforts will be embedded must be examined. This involves identifying all potential participants in the process, their particular perspectives and values, the situations in which they operate, the strategies they use to achieve desired outcomes, and the effects of these outcomes. An understanding of the historical and cultural trends that have influenced the roles played by participants in the process, as well as the factors that condition these trends, helps in clarifying the common interest and achieving recovery goals.

The wolf is one of the most studied large mammals in North America. Thus a great deal of information is available to scientists and managers charged with restoring wolves to former habitats. For example, the recently completed study of the feasibility of wolf reintroduction to the Olympic Peninsula in Washington State consulted nearly 1,000 relevant documents in an effort to be comprehensive yet selective (Ratti et al. 1999).

The Rocky Mountain Wolf Recovery team was formed in 1975 to draft a workable recovery plan, marking the beginning of a 15-year period of intelligence gathering. During this period, biologists searched for evidence of the presence of wolves in the GYE, found none, and recommended reintroduction. Educational and promotional efforts, such as the Science Museum of Minnesota's "Wolves and Humans" exhibit, were launched and generated interest in and support for recovery. Political allies, such as Utah Democratic Representative Wayne Owens and Idaho Republican Senator Jim McClure, were recruited after considerable efforts on the part of key individuals and organizations. As a result, legislation that directed further study of potential impacts was introduced with the caveat that interests of agricultural constituencies would be protected. In a good faith effort to encourage support for recovery among livestock producers, Defenders of Wildlife established a $100,000 wolf compensation fund that promised to reimburse ranchers for any livestock losses. Near the end of this stage of the program, a reintroduction plan was developed by the Wolf Management Committee as directed by the secretary of the interior.

Much information was generated to determine potential impacts of wolf reintroduction and make decisions about how to proceed, but the emotional and often hostile debate that ensued indicated continued resistance to the idea of recovery at local and regional levels. People exhibited strong feelings toward wolves, either for or against, and there was little middle ground (Huffman 1993). Some maintained that the debate represented resistance to social change and outside agendas rather than opposition to wolves themselves. For example, Renee Askins, formerly of the Wolf Fund, equated the return of wolves to relinquishing control of the West. Wolves become a symbol of resulting painful changes, with their associated sacrifices and compromises (Askins 1993). In this atmosphere, conflicts were inevitable and would later prove problematic.

Lesson Three

Lesson three focuses on the selected course of action to achieve recovery goals after all available options have been formulated, debated, and evaluated. The choice of the preferred alternative stabilizes expectations by clarifying information, rules, and implications for recovery. The traditional approach of preparing an environmental impact statement (EIS), which evaluates a number of alternatives and identifies the preferred option, may not be sufficient to legitimize the process and gain adequate popular and political support. More attention to social context, specifically the recruitment of respected local and regional elites, may be necessary to encourage endorsement of the prescription.

In comparison to the previous phase of estimation, this stage of the recovery program progressed rapidly. The Wolf Management Committee submitted its recommendations to Congress in 1991, a year that also witnessed the funding of an EIS. Information was again gathered and reviewed to evaluate reintroduction of an experimental, nonessential population as a means of wolf recovery and to identify and consider all possible alternatives. Numerous public hearings were held and more than 160,000 written comments were received, the most ever generated by a federal action. The majority of opinions offered were favorable to the preferred alternative. During this time, Defenders of Wildlife continued to promote recovery by staffing "Vote Wolf!" booths in Yellowstone National Park. They collected more than 70,000 ballots, nearly all of which registered support for restoration. However, strong opposition still remained, and in 1994 the Wyoming Farm Bureau filed a lawsuit indicating its adversarial stance.

This lawsuit, when combined with one filed by a Wyoming couple and another filed by a coalition of environmental groups, would later result in the decision to remove reintroduced wolves from the GYE. The suit is under appeal and illustrates the dissension among participants in the recovery process. Environmentalists disagreed among themselves about the method of recovery to be promoted. Some preferred natural recovery to reintroduction, pointing to the close proximity of the population of wolves in northern Montana and the presence of individual dispersers in Yellowstone. It was only a matter of time before more wolves made their way south and established a population on their own. Natural recovery, proponents maintained, was more acceptable to local residents who react negatively to government interference, particularly in the form of costly programs (Cromley 1997). Additionally, wolves would enjoy full protection under the ESA rather than being considered experimental and, in this case, nonessential.

Those in support of reintroduction argued that the management flexibility allowed with such an experimental designation and the resulting compromise in protection was essential to gaining local acceptance. They pointed to statements made by David Mech (1995b, E6), a recognized expert in wolf ecology, who concluded that the "best way to ensure recovery is to not protect [wolves] completely." They also reminded their colleagues that decades passed before reproduction was documented in Montana and a population of wolves was able to establish itself. Natural colonization, according to Hank Fischer of Defenders of Wildlife, was an iffy proposition at best (Hacket 1993). Reintroduction, proponents claimed, offered an expedient and realistic alternative to natural recovery that could also avoid the potential for genetic bottlenecks. Despite warnings by environmentalists and livestock producers that a decision to reintroduce wolves would result in legal challenges, the prescription was made and questions were raised in the minds of interested observers.

Lesson Four

Lesson four stems from the implementation of the selected course of action. Rules are interpreted, supplemented, and enforced to achieve recovery goals. The program's relationship to existing institutions is defined, as are incentive structures. Costs are minimized and performance expectations are detailed. This stage of the program, in which wolf recovery in the GYE is currently entrenched, requires consensus and cooperation between participants in the process to ensure success. Technical considerations certainly are important, but emphasis on the education and involvement of sympathetic members of the opposition, especially at local and state levels, can limit hostility and reduce potential conflicts. Without local acceptance, however grudging, recovery can be compromised.

Once the preferred alternative, namely reintroduction as an experimental, nonessential population, was approved by the U.S. Fish and Wildlife Service and endorsed by the secretary of the interior, implementation proceeded in the GYE. This entailed capturing and translocating wolves from Canada to Yellowstone and central Idaho in 1995 and 1996, and subsequent monitoring of the reintroduced populations. The wolves adapted readily to their new environments. Pairs formed and established territories, and populations burgeoned. In fact, breeding was so successful that a third planned input of wolves from Canada was no longer deemed necessary to assist with recovery. Indeed, it appeared that recovery goals would be met earlier than expected.

Although wolves responded favorably to reintroduction, local residents, particularly livestock producers, were skeptical of promises made during program development. It was important that managers kept their promises to alleviate their fears and mistrust of government. Incidents of depredation were investigated and dealt with as quickly as possible, and ranchers were compensated for the market value of confirmed losses to wolves. However, as wolves ranged beyond the boundaries of protected areas, fears of restrictions on the use of public lands and an increase in federal control became more pronounced and the recovery program less tolerable. Wolves found outside Yellowstone National Park and other protected areas exist in a hostile climate. To ranchers, wolves heralded unwelcome changes in the traditional way of life in the West, with "outsiders dictating agendas" (Mader 1993). Their response? Kill the messenger.

Lesson Five

Lesson five focuses on appraisal of the program, appraisal that examines earlier established goals and all preceding program activity with a particular emphasis on the success or failure of implementation. A comparison is made between estimated performance levels and those actually obtained, and quality of performance is assessed. Evaluation can be formal or informal and can include both internal and external review. Decisions about policy prescriptions are scrutinized, and responsibility and accountability for the outcomes of those decisions are appraised. Information is obtained, and recommendations about how to proceed are disseminated appropriately.

Evaluating program success should not be limited to biological aspects but should include an ongoing assessment of the social ramifications. Increasing numbers of wolves may be a biological indicator of success, but strong trends toward renewed opposition to recovery may have much greater consequences for the long-term success of restoration efforts.

A cursory evaluation of the recovery program reveals better than expected success in achieving recovery goals. Wolves are breeding and rapidly approaching change in their protected status and future management. They are fulfilling their role as top predator and restoring ecological balance to the GYE, already revealing insights into predator–prey dynamics and their indirect effects. Incidents of depredation are managed according to prescription by prompt control of offenders and direct compensation for losses. Information about individual wolves and packs is updated and distributed regularly to interested parties. Visitors to Yellowstone National Park are rewarded with sightings of wolves far beyond what was predicted, and their keen interest in the recovery effort has inspired and supported the proliferation of wolf books and other paraphernalia.

Upon closer examination, however, a number of problems (both real and potential) become evident. For example, the accelerated rate of recovery has stepped up plans for delisting and transfer of management responsibilities to state agencies. For reasons explained later in this chapter, this may present difficulties. In addition, managers of the recovery effort can no longer monitor the whereabouts of every wolf, and ranchers are not pleased. Incidents of depredation, no matter how few, fuel their antipathy. Compensation for losses is welcomed but is paid only for confirmed wolf kills and is considered by livestock owners to be insufficient to cover indirect costs of harassment (e.g., stress, low birth weights, trampling of pastures). Questions still remain about whether losses will be compensated after wolves are delisted.

Although the recovery program represents a substantial effort to accommodate local interests and ward off potential conflicts, these examples and continued litigation illustrate the need to fully attend to the social context of all such programs. Short-term successes do not necessarily ensure long-term viability of recovered populations.

Lesson Six

Lesson six concerns termination or redirection of the restoration program. This stage represents an opportunity to stop or modify practices that are not working and those that have accomplished their goals and move forward to a new beginning. Endings or transitions inevitably occur during the course of a program. Changing the protected status from endangered or threatened to recovered, for example, is one goal of endangered species restoration programs. However, many such programs fail to prepare for termination early in their development, which can result in difficulties. Delisting wolves in the GYE, accompanied by the transfer of management authority to the states, is a case in point (Clark 1996b).

Early attention should be paid to this stage of program development to anticipate potential areas of conflict that could reduce chances for successful recovery. It is also important to remember that changes in program direction or management can be difficult because participants often are emotionally invested in their roles in its development and committed to pursuing a particular course of action. Sensitivity to such difficulties is important.

Responsibility for implementing and managing wolf recovery in Yellowstone National Park and central Idaho rests primarily with the U.S. Fish and Wildlife Service, although the National Park Service and Nez Perce Tribe play important collaborative roles. The population of wolves in the park is fast approaching the 100 animals needed to qualify for recovery and delisting. Still, at least 10 breeding pairs must be established for 3 consecutive years in each of the recovery areas after reintroduction to meet accepted criteria.

In discussions of the transfer of management authority and control from federal to state agencies, the states of Montana, Wyoming, and Idaho are reluctant to accept what they perceive as a substantial management expense. The agencies responsible for managing state wildlife populations want to exercise their authority and control, but they are unwilling to commit their limited resources to maintaining a recovered wolf population without a promise of permanent funding from the federal government. The power and role of the states relative to the federal government is a major issue affecting wildlife management and the implementation of federal endangered species policy (Clark

1997a). It stems from the claim that with the creation of the ESA the federal government usurped states' ownership of resident wildlife and thus their traditional management authority and control. The current manifestation of the states' rights ideology highlights the sociopolitical forces that inhibit clarification of the common interest and undermine cooperative efforts to promote wolf recovery. Western intolerance of federal involvement in what is perceived to be a state issue is growing. It remains to be seen whether reintroduced wolves will survive the struggle for management authority and control of public lands.

Recommendations for Improving the Process

Wolf restoration in the GYE is a monumental achievement in endangered species recovery. It has been hailed as a model of success. Certainly, wolves would not argue with that assessment because their numbers are rapidly approaching recovery goals. But despite such indicators of success, the effort has failed to win the acceptance and support of many powerful local residents on which the long-term future of wolves in the GYE ultimately depends. It is apparent from continued conflicts that more attention must be paid to social impacts, both early in program development and during the later stages. Our approach to endangered species recovery may be greatly improved by incorporating the following recommendations.

Be Problem Oriented

To navigate through the emotional and political morass of increasingly complex conservation problems with some hope of arriving at long-term solutions in the common interest, it is necessary to become problem oriented (Wallace and Clark 1999). By focusing attention on the problem itself, it can be analyzed in relation to its entire context and approached from the standpoint of rational awareness. A number of intellectual tasks involved in this strategy can be used to define problems more clearly and craft effective solutions (Clark et al. 1996a). The first task is to clarify the goals of the participants and determine the range of perspectives and values held in relation to the problem. The second and third tasks require developing a thorough understanding of the historical trends that have influenced the nature of the problem, including the conditions under which these trends have evolved. Once the historical context is understood, the fourth and fifth tasks may be carried out whereby future trends in the problem are projected and analyzed, and alternatives are invented and evaluated as possible solutions. It is important to remember, however, that creative problem solving is largely an iterative process. Expect to revisit these tasks to refine chosen solutions.

Be Contextual

The importance of the human factor in endangered species recovery cannot be overstated. Mapping the social context in which conservation problems such as wolf reintroduction are embedded is essential to distilling the debate and promoting tolerance and understanding among participants. People tend to behave in ways that they perceive will leave them better off than if they had acted differently. Perceptions differ between people, and these differences may be vast and seemingly irreconcilable. However, by attending to people's perspectives and values, human dynamics and their implications for wolf recovery may be better understood. The social context is not limited to individual participants but includes the institutions through which people pursue their values. Insights gained through its examination can suggest practical improvements and ways to engender support for recovery efforts. The social process, by definition, is not static and should be remapped continually over the life cycle of any endangered species recovery program to anticipate conflicts and respond with creative solutions that encourage public participation (Clark and Wallace 1998).

Be Common-Interest Focused

Wolf recovery is a human endeavor and therefore fraught with all the difficulties that arise when attempting to clarify and secure the common interest among diverse participants. If the social context reflects a collection of narrow self-interests, then the stages in the life cycle of a recovery program represent an opportunity to reconcile these differences and foster development of the common interest. A working specification of the common interest generally takes the form of rules (decisions). Rules are necessary to coordinate efforts between participants and make informed decisions that meet with expectations outlined in recovery plans, cooperative agreements, and other such documents. However, the existence of rules does not preclude participants from pursuing their own special interests at the expense of the common interest. Accountability is needed to maintain respect for agreed-upon rules of conduct and the quality of the program itself. Participants familiar with the decision process, how it works, and how it can be monitored are better able to intervene and improve decisions made and, in effect, support the partnership that has been formed. An open, flexible process that encourages mutual exchange can minimize destructive conflict and secure the common ground that effectively solidifies partnerships and ensures success (Clark and Brunner 1996).

Conclusions

Working to improve the way carnivore conservation is understood and approached can foster management efforts that are more efficient and effective at both local and regional levels. In turn, the process improvements we promote can lead to more favorable conservation outcomes.

Human values and attitudes toward wolves have changed with time and with our increased knowledge and understanding of the role predators play in natural ecosystems, the importance of which has been displayed dramatically in the ecological changes that have occurred in the brief time since wolves were reintroduced to Yellowstone National Park. Thousands of visitors have borne witness. However, the complex historical relationship between wolves and humans undoubtedly continues to influence wolf recovery. In particular, the conservative climate that settled and shaped the West and was largely responsible for extirpating wolves from the region and throughout the contiguous United States still maintains its hold in prevailing myth and undermines attempts at reconciliation with the emerging environmental ethic. To achieve successful reintroduction and ensure establishment of wolves in the GYE, as well as their responsible future management in the common interest, a system approach is recommended. Such a holistic approach considers social acceptability to be as important to the long-term success of recovery programs as the numerous biological and technical factors.

The world is watching. How we choose to proceed in the GYE may have direct consequences for future recovery efforts, such as the proposed reintroduction of wolves to the Adirondacks. Certainly, a contextual, problem-solving approach that promotes cooperative inquiry within an adversarial context and encourages participants to engage in reflective conversation about their values and attitudes offers the greatest opportunity to improve the process and achieve the common interest.

Acknowledgments

Peyton Curlee Griffin critically reviewed the manuscript. Conversations and joint analytic exercises with colleagues, former students, and diverse professionals from more than 25 countries during the last 15 years also greatly improved its quality. Support to produce this chapter came from the Hastings Center, the American Museum of Natural History, Denver Zoological Foundation, Cathy Patrick, Judy Gould, Gil Ordway, Fanwood Foundation, NewLand Foundation, Northern Rockies Conservation Cooperative, and Yale University.

Endnote

1. The district court consolidated the challenges brought by both the Farm Bureau and environmental organizations and found that the rules used to implement the Yellowstone reintroduction were contrary to the intent of the Endangered Species Act. The court stayed its own judgment pending appeal. On January 13, 2000, the U.S. Court of Appeals 10th District reversed the district court ruling (see *Wyoming Farm Bureau v. Babbitt* 2000).

Chapter Twelve

Wolves as Top Carnivores: New Faces in New Places

Rolf O. Peterson

In modern geological times, the gray wolf *(Canis lupus)* was, with the exception of the red fox *(Vulpes vulpes)* and our own species, the most widely distributed mammal in the northern hemisphere. It plays an ecological role shaped by evolutionary pressures that predate the appearance of the human species. Yet serious efforts to understand the complexities of wolf relationships with other species span just a few short decades in the last half of the twentieth century. The elimination of wolves and other large carnivores probably had profound ecological impacts about which we can only guess, but the resilience of ecosystems and the species that remained brought quick adjustments, often guided by large-scale agricultural interests of pioneering human societies. Restoring wolves to some of their former haunts may once again imply substantial future adjustments in ecosystems and human relationships to wildlands because we are restoring an ecologically important large carnivore that occupies a large carnivore niche that we have not completely vacated ourselves. In this chapter I attempt to describe the wolf in its ecological and evolutionary contexts, using examples drawn especially from my long-term studies of wolves in Isle Royale National Park, and provide recent perspective based on the return of wolves to Michigan's Upper Peninsula.

The Ecological Role of the Top Predator in the Food Chain

Wolves were almost eliminated in the continental United States because of pervasive human fear of this carnivore among European immigrants and because the wolf competed directly for domestic prey introduced by newly arrived human settlers. These historical realities tell us little about the real wolf that is being actively restored by its own initiative or by wildlife managers.

Field studies of wolves in the 1930s and 1940s by Sigurd Olson in Minnesota and Adolph Murie in Alaska began to provide a new perspective on the ecological role of the wolf, an endeavor that continues to this day in many locales around the world. More general ecological insights have developed that emphasized the importance of food shortage in regulating animal populations (Lack 1954), and new research in the 1950s and 1960s illustrated that wolf predation was highly selective. Popular books about wolves (Crisler 1958; Mowat 1963) spread the notion that highly social wolves were both fascinating and benign in their ecological role as predator, despite early suggestions that wolves probably limited prey populations (Mech 1966; Pimlott 1967).

Through the 1970s wolves began to receive mixed reviews as they were increasingly implicated in prey declines. Some studies (Caughley 1977; Peterson 1977; Houston 1982) continued to emphasize habitat availability and food supply in regulating prey populations; these are often called density-dependent mechanisms because the dampening effects of resource shortages become more acute as population density increases. Other studies pointed to major declines of prey that may have been initiated by overhunting but were greatly extended by wolf and bear predation (Gasaway et al. 1983).

Wildlife managers discussed whether food or predators limited game populations, whereas theoretical ecologists debated the same question: whether trophic systems in ecological communities were structured by bottom-up influences of resource limitation or top-down effects of consumers. The view that populations were regulated by density-dependent mechanisms was countered by the notion that density-independent forces such as the weather were of paramount importance. More elaborate ideas evolved into food chain theory, provocatively set forth (Hairston et al. 1960) as an explanation for why the world is green: Herbivores are strictly limited by carnivores, thereby sparing green vegetation. Forty years later this simple hypothesis stands as a good first approximation of the ecological role of the wolf. Let's attempt to apply it to the specifics.

Effects of Wolves on Ungulate Prey Density

At a very gross level, we can say a few things about what wolves might do in an ecosystem. Most importantly, wolves probably will reduce ungulate or hoofed animal abundance. Not always, and not in all places, but because not every animal wolves kill would otherwise die soon from other causes, we can assume that wolf predation is not completely compensatory. Mech (1995c) stated that if you "remove the wolf, prey numbers will generally increase" (538); the logical corollary is that prey will decline when wolves are restored to a place from which they were extirpated. It is difficult to provide a more detailed prediction, even for a specific system such as the Adirondacks, because wolves would rejoin an ecosystem with complex interactions and dependencies between resident species, where alternative prey, multiple predators, time lags, severe winters, and parasites and diseases, not to mention human actions, all come to bear on outcomes, which vary over time and space. Given the range and duration of our experience (i.e., state of our science), such complexity precludes predictability. Consequently, it is best to view any wolf introduction as an experiment and consider all predictions as possible hypotheses.

Nevertheless, the presence of large carnivores does affect prey populations. The best available data come from moose and caribou, the easiest prey populations to enumerate (Figure 12.1). Considering that wolves, bears, and humans are all important top carnivores, it can readily be shown that with each addition of a predator species, there is usually an associated decline in prey density. Where only one predator is present, prey tend to be abundant. If a second predator species is added, ungulate abundance goes down. If all three predators are present—wolves, bears, and humans—ungulate abundance usually is quite low.

The presence or absence of wolves does appear to influence long-term average densities of their prey. Given the presence of wolves, however, the fluctuations in prey noticed by people will arise largely from factors other than wolf predation; in the context of the Adirondacks, winter severity is the primary forcing agent. Also, wolf predation might vary because of disease among the wolves or the presence of alternative prey. Likewise, alternative prey species might affect the influence of wolf predation on a central prey species.

Although the presence or absence of predation may determine average prey abundance, the short-term fluctuations in prey that are of interest to most people usually are caused by other factors. That is, prey numbers usually are

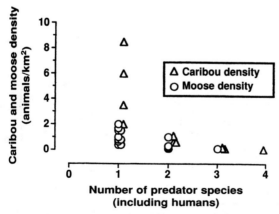

Figure 12.1. Density of moose (O) and caribou (△) in relation to the number of coexisting predators, based on 30 studies of moose and caribou in the northern hemisphere (data from sources cited in Peterson 1999 and Caughley and Sinclair 1994).

correlated not with wolf numbers but with other factors (e.g., winter severity) that produce disturbance in the system (Boyce and Anderson 1998).

To the basic predator–prey equation we must add many subtleties about which little can be said because our science is usually short-term, our measurements are crude, and outcomes may be case specific. One locale in which some teasing apart of complexity has been possible is Isle Royale National Park, a 544-square-kilometer wilderness island in Lake Superior. Isle Royale provides a simple wolf–prey system in which wolves rely on moose and secondarily on beavers, and immigration and emigration do not complicate counts of animal populations. Here there is no hunting by humans. Mech (1966) found that wolves killed primarily juvenile or very old moose, leaving the breeding population of moose largely intact. Wolf predation on the old moose is inconsequential because many would soon die of other causes (compensatory predation). That is to say, wolf predation on superannuated moose is compensatory. However, predation on young animals that have not yet reproduced is a different matter because these animals will determine the future trajectory of the prey populations, and many of these young prey would not otherwise perish so soon (additive predation).

Prey populations will continue to fluctuate, with or without wolves, as they always have. The balance of nature is not static. At Isle Royale these fluctuations appear to recur about every 20 years (Figure 12.2) in a manner that resembles the classic predator–prey cycles between snowshoe hares and lynx. Although these fluctuations may be more evident at Isle Royale because it is such a simple system, it is worth noting in passing that population fluctuations in the moose of Isle Royale often have been mirrored in many mainland populations across North America. As usually seen in predator–prey cycles in smaller mammals, the increase in predators tends to follow increases in prey,

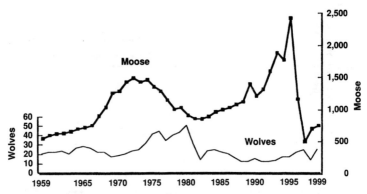

Figure 12.2. Abundance of wolves and moose in Isle Royale National Park, 1959–1999 (from Peterson et al. 1998 and unpublished data 1999).

illustrating the obvious dependency of top carnivores on their prey. At Isle Royale, moose more than 10 years old are the class of prey that feeds the wolves, explaining the very long lag between increases in the moose population (from increased survival of calves) and subsequent increases in wolves.

The significance of density-dependent responses as prey populations reach high levels is a point of interest, given the prominence of this concept in population dynamics theory and its significance in the natural regulation of wildlife populations in national parks (Peterson et al. 1998). By this notion, reduced nutrition that accompanies high numbers causes reduced reproduction, thereby stabilizing the population. It is clear that reduced nutrition is prevalent at high population densities; this is evident in shorter leg bones and longer period of growth for Isle Royale moose born at high population densities. However, calf survival did not decline until something tipped the balance and gave wolves the upper hand. Increased predation, not reduced reproductive success, brought the moose population down from its peak in the 1970s.

Even more telling was the response of the moose population to reduced wolf numbers in the 1980s, after lethal canine parvovirus arrived and the wolf population crashed to its lowest level. Moose simply grew in number, despite the nutritional distress evident in the bones of growth-retarded moose calves. Nothing stopped the growth of the moose population, except an outbreak of winter ticks in 1989 (DelGuidice et al. 1997), until the unprecedented severity of the winter of 1995–1996 caused massive starvation losses and the moose population crashed. Contrary to my earlier conclusion (Peterson 1977), the moose of Isle Royale do not exhibit any significant tendency to self-regulate.

Wolves and Beavers

As the fur trapper rapidly becomes a historical figure in America, the species that was the trapper's target, the beaver *(Castor canadensis)*, is thriving to an extent that no one alive has ever witnessed (Williams 1999). In upper Michigan, where 15,000 to 20,000 beavers were once trapped annually, the take by trappers has been reduced by 75%. Beavers, it seems, are intent on reengineering the entire landscape, including a small creek just two blocks from my house in the small city of Houghton, Michigan. The beaver is also proliferating in upstate New York and is therefore a potential prey of any wolves that might come to inhabit the Adirondacks.

Isle Royale may be the only locale where numbers of beavers and wolves have been counted for enough years to provide a few clues about their interaction. When wolves reached peak population size in 1980, the beaver population plummeted (Figure 12.3). As wolves themselves declined in the early 1980s, the beaver staged a remarkable comeback, even in the old forests that

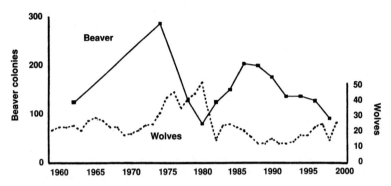

Figure 12.3. Beaver abundance in Isle Royale National Park in relation to wolf population size (data from Peterson et al. 1998 and D. W. Smith, unpublished data 1999).

have gradually lost their appeal to this industrious mammal. Beaver are a highly prized prey species for wolves because they don't defend themselves nearly as well as most ungulates, and wolves must value highly the large store of fat available in a beaver carcass. Based on the Isle Royale example, when wolves exist at high density they can apparently limit beaver numbers. Although wolves cannot survive year-round on beavers, which are locked under ice in winter, beavers nevertheless provide an important source of food for wolves during the open-water season.

Wolves and Other Carnivores

An important implication of the food chain hypothesis (Hairston et al. 1960) is that top carnivores are regulated by competition for food. We should expect to see evidence that the largest and most powerful carnivores might attempt to displace or otherwise get rid of smaller carnivores. Pervasive evidence of such competition among wild canids has become clear over the past two decades (Peterson 1995). It is manifested not just in squabbles over food near wolf kills, although these are common, but in everyday behavior of larger canids toward smaller cousins. Particularly in winter, wolves often chase down and kill smaller canids. As long as the cost of doing so is small, theory predicts that wolves should behave in this way, ridding their world of anything that might eat their food. Indeed, we are no different. Human intolerance toward other large carnivores is a simple expression of the ecological reality of competition.

The pervasive significance of competition has also shaped wolf behavior such that our beloved dogs, transformed from wolves by our own ingenuity, are at risk in wolf country. The propensity of wolves to kill dogs is an act not of predation (direct food acquisition) but rather of competition. The evidence of this is that dog carcasses are often left uneaten by wolves. Such behavior does little to endear wolves to rural residents, but this evolved behavior comes along with the real-world package we call the gray wolf, as does the historical tendency of wolves to kill domestic stock as prey.

Cascading Influences of Wolves

What of the ecosystem implications of wolf predation? If wolves, together with other carnivores (including hunting humans), limit prey populations, does this influence cascade through the ecosystem to influence the plants that support the prey? In the simple system at Isle Royale, with one major predator and one primary prey, it has been shown that the plants indeed fluctuate in concert with the wolf population. When wolves are numerous, trees grow (McLaren and Peterson 1994). By extension, anything that depends on trees growing or not growing must also be affected indirectly by the wolves.

In human-dominated ecosystems, even some of the wildlands where wolves may become reestablished, the overall significance of wolves as an ecosystem player may be muted by human effects. For example, wolves may avoid areas with high human populations, humans may kill wolves by cars and by legal and illegal killing, or humans may benefit prey by providing food supplements for deer, either directly or through agriculture.

Coevolution of Wolves and Prey

It is evident that wolf restoration is not simply the return of the wolf. It also means the return of everything that wolves do, directly and indirectly, in an ecosystem. Although these are some of the short-term implications for modern-day humans, it should be appreciated that wolves and other large carnivores were part of the fabric of evolution that shaped big game animals. The huge size of the moose was established during the Pleistocene when the giant short-faced bear, a carnivore as tall as a moose, roamed the northern hemisphere (Geist 1997). The large body size of moose may have developed in part as a defense against a carnivore that is now extinct.

One argument in favor of wolf restoration is that we are perpetuating the forces of natural selection that led to the development of big game species as we know them today. Long-term maintenance of species characteristics probably depends to a degree on continuation of similar selective forces, but this is unlikely to be evident on anything less than a geologic time scale. However, it may not be widely appreciated how prey behavior represents adaptations to living with wolves.

Nelson and Mech (1981) argued that winter yarding behavior in white-tailed deer, familiar in the Adirondacks and from Maine to Minnesota, was not a means of seeking favorable thermal cover as much as an evolved behavioral tactic to reduce the risk of dying of wolf attack. As winter snow accumulates, deer and other cervids are more at risk from wolves, and they congregate in groups (safer than as singles) where snow is less deep. It can still be claimed that the thermal cover is important for deer in winter, but the antipredator hypothesis also would explain similar wintertime movements to coniferous cover also exhibited by moose (Peterson and Page 1993), so large in body size that thermal cover is unnecessary.

That alert senses, fleetness of foot, and precocial young may have evolved as antipredator defense seems straightforward, but the influence of eons of wolf predation may go deeper into subtleties of social behavior. The survival pattern of moose on Isle Royale in relation to age, summarized in life tables (Peterson 1977), for example, has been well documented. Wolves often kill moose that are less than a year old, but beyond that vulnerable period moose grow up to prime breeding age (5 to 10 years for males, perhaps 3 to 15 years for females) without much risk from wolves; they are able to protect themselves effectively by standing their ground (Mech 1966). Old moose develop various diseases and disabilities that render them vulnerable to wolves (Peterson 1977). This ages-old pattern of mortality has implications for evolved patterns of social behavior, in which, for example, competition between males for access to females and female choice of males is correlated with antler size and age. In other words, social status is highly dependent on age and vigor, and social balance (and greatest reproductive success) in a moose population is best achieved by having an ample proportion of prime breeding animals in relation to younger, less mature individuals (Bubenik 1972). Shortly before his death, ethologist A. B. Bubenik determined from simulation modeling that the population age structure for moose that most closely approximated his notion of social balance was the same as the long-term age structure of Isle Royale moose, managed by wolves, not the rifles of modern hunters. Improvements in moose populations have accompanied recent efforts to protect prime breeding animals (Schwartz et al. 1992).

The preceding comments on the evolutionary significance of wolf predation in cervid ecology and behavior do not imply that we must have wolves present everywhere to manage deer and other prey species properly. The modern hunter, properly managed, can do a remarkably efficient and effective job of curtailing population increase and stabilizing cervid populations to achieve management objectives that are compatible with human societies. On the other hand, where it is possible to retain large carnivores together with prey, would it not be prudent to do so, in accord with the Leopold dictate to "save all the pieces" while trying to maintain wildland resources? Literary recognition of the evolutionary significance of wolves was penned long ago by Robinson Jeffers (in Lopez 1978, 269):

> What but the wolf's tooth whittled so fine
> The fleet limbs of the antelope?

Public Sentiment About Wolf Reintroduction

It has been 8 years since wolves were confirmed as reproducing successfully in upper Michigan, after an absence of almost 40 years, and it would not be an exaggeration to say that few residents have even taken notice. Of course, this is likely to change as wolves increase in number and distribution to fill their former niche. The greatest future challenge may be to decide, as a society, where wolves will be protected and tolerated and where they will not and the methods by which this balance between humans and wolves will be achieved (Mech 1996).

Lack of public concern, even interest, contrasts sharply with the strong reaction evident 25 years ago when four wolves from Minnesota were released in upper Michigan, prompting a flood of protest (Weise et al. 1975). In part, this reflects generational change in attitudes about wolves, but public surveys in Michigan also revealed that wolves would be more tolerated if they arrived on their own rather than through reintroduction by government agency (Kellert and HBRS 1990). Finally, the group that consistently has the lowest tolerance for wolves—farmers—has been disappearing steadily from the landscape of upper Michigan over the past 25 years. These three factors all seem significant in shaping local attitudes toward wolves.

For 3 years in the mid-1990s I helped distribute information about wolves to deer hunters on the day before the opening of the firearm deer-hunting season and, in so doing, talked to about 300 deer hunters. Consistent with published attitudes surveys (Kellert and HBRS 1990), I found that deer hunters were very favorably disposed toward wolves. They appear to place high value on an experience in the wilds of upper Michigan, and the wolf adds to the aura of the total experience.

Conclusion

Through overt introduction and increased tolerance, we have returned the wolf to a few wildlands in North America where it formerly existed. This represents a sea change in human attitudes toward a controversial species. It is important for people to understand the actual role of wolves in wild ecosystems. As a top carnivore, the wolf exerts an influence on other species and on the ecosystem that is disproportionately large compared to its numbers. We can anticipate that prey populations and some competing species will be reduced in regions where wolves become firmly reestablished; this outcome will be viewed positively or negatively by different groups, depending on knowledge, experience, and attitudes. The wolf, long an agent of change for prey species throughout their evolutionary history, will once again begin to fill its ancient niche. Its presence may be manifested throughout a food chain, primarily through the reduction of certain prey species. Our ability to predict the outcome of wolf reestablishment is low, particularly the future responses of human societies, which will continue to decide where wolves will be allowed to live.

Chapter Thirteen

Reintroduction: Inspired Policy or Poor Conservation?

Christine L. Schadler

A blue, iridescent curtain of shimmering northern light undulates in the sky while across the land, a caravan of double headlights twenty kilometers long snakes its way through the Canadian night. It's August, and 2,000 people have come to hear wolves howl.

The continent's oldest orchestrated wolf howl at Algonquin Provincial Park in Ontario draws mothers and fathers with their children, a group from England, college students backpacking across Canada, and tourists of all description who have come by car from as far away as Detroit and Montreal to witness the primeval wilderness. The caravan pulls to a slow stop near a wooded bog where wolves were heard the night before. We hope to hear them, but understand there are no guarantees in nature— the wolves may have moved on.

We have been instructed to remain absolutely quiet. A thousand muffled doors closing indicate we have emerged from our cars and the stillness signals our tense readiness. The man howls. Twice. He sounds like a siren. Wolves can't resist a good siren! In a few seconds we hear a whine, then a whimper-cry, a few more join in and, gaining momentum, the pack choruses in with crooning cadence, yips, yipes, squeals, yaps, barks, a rolling a cappella, dinner music predator-style, replete with "wagging tails and playful maulings," as Leopold (1949, 129–30) describes, all hidden from view, of course, by the night.

When they stop, the stillness of the Canadian night closes in around us once more. I hear the throaty exhalations of those around me and see smiling faces beaming in the starlit night. Everyone is thrilled! "That was the coolest thing I've ever heard," said one college student. "I have peered into the primeval and it was glorious," says a British gent. Many of these people traveled hours to experience a minute of howls, but said it was worth it.

We climb back into our cars and the caravan slowly wends its way back to civilization. I wonder how the experience would change if everyone had been instructed to place their ticket in a basket and at the end of the evening, one name would be drawn—the lucky winner would be granted the opportunity to spend the night camped—solo—by that wooded bog and experience the wolves and the sky and the land, one on. How different would I feel knowing that I might be the one to forsake the safety of my trappings and occupy a place in the wild—just another creature—and not at the top of the food chain, either! What would be the meaning, then, to me, of the howl of the wolf? I wondered, as I considered this, about the real meaning of not only these wolf howling nights at Algonquin, but of our current interest in wolf reintroductions.

Does the thrilled response of these thousands of people indicate an essential, fundamental change in our relationship with wild animals? Short of being dropped off by the bog with a tent and some victuals, are we any the more likely to tolerate wolves living on the far side of the field? Are these wolf howl nights and our reintroduction efforts a clear indication that parity between humans and the wild is a nearer reality? I think not. I think there is danger in believing that a reintroduction becomes a panacea that can undo 2 centuries of wrongs and 20 centuries of antipathy toward nature. We lull ourselves into believing that by reintroducing wolves into wild parklands, balance will once again be achieved or even that reintroduction programs constitute good long-term environmental policy.

The most recent proposal to reintroduce wolves into Adirondack State Park in New York and into the North Woods of New England provides a template on which to examine the assumptions inherent in manipulating nature and the social and biological fault lines in current wildlife management thinking. In this chapter I examine the relative merits of reintroduction and natural restoration, drawing on the natural recovery of wolves to the Great Lakes states. In support of an argument for natural recovery, I discuss the taxonomic difficulties of identifying the wolf subspecies formerly inhabiting the northeastern United States and reflect on the curious disparity in our society's view of the coyote and the wolf.

Diverse Perspectives on Wildlife Policy

The views of the American public are heterogeneous with regard to wildlife in general and wolf restoration. The attitudinal differences between rural and urban dwellers are exacerbated by the fact that the latter group is increasingly deciding wildlife policy in this country.

Most Americans live in cities or suburbs, where issues such as a wolf reintroduction seem distant and are experienced only remotely, through TV or newspapers. These millions of people constitute the broad majority of Americans who, living removed from wilderness, are increasingly supporting conservation initiatives by donating money to organizations or responding to surveys asking for their opinions and support. Recently, when conservation organizations have taken the American pulse on wolf reintroductions, they have asked urban Americans questions such as "Should wolves be returned to their rightful place—say, in Yellowstone?"

Of course, this sounds reasonable and right, so the "Yes" box is checked. Or, "Would you support reintroducing the wolf if it meant the ecosystem could, once again, be complete?" Surely, how could this be wrong? When these are taken as essentially moral questions, no one wants to check the wrong box. The results of these surveys inform policymakers that an overwhelming majority of Americans appear to favor wolf reintroduction.

However, these urban Americans reside hundreds and sometimes thousands of miles from any wolf or potential wolf habitat. The idea of the wolf, wild and free, remains sanctified and safe. The respondent, having played a part by checking boxes or paying tithe, nurtures a newfound goodwill by participating in environmental restoration, having very little knowledge about the political, economic, or biological ramifications of a reintroduction.

Meanwhile, hundreds of miles away, within earshot of the howl, live people who feel overlooked and resentful of yet another federal program. They have attended public hearings and voiced their opposition to reintroduction proposals. Many have angrily declared their intention to thwart reintroduction efforts. To mollify these citizens and to acknowledge their economic and other interests, Section 10(j) of the Endangered Species Act (ESA; 16 USC 1539j) has been invoked, which offers local land users (ranchers and farmers) the chance to "manage" (kill) wolves when they cross park boundary lines into farmed land, whether it be publicly or privately owned.

Back in the city, good-intentioned people don't learn about or understand the phrase "experimental, nonessential," which is Section 10(j) of the ESA. In my experience, when they learn the ramifications of this compromise, they feel discouraged. For instance, when they learn that, given time, wolves may have returned on their own to Yellowstone and were recovering along the chain of the Rockies with the full protection of the ESA, they wonder why this reintroduction was necessary. For many Americans, lack of awareness and remoteness from the issue translates eventually into lack of concern.

With the passage of the ESA in 1973, we declared our intention to rectify a past of reckless disregard for wildlife and to recognize that species and ecosystems had been imperiled by that disregard. Implicit in the passage of this act is the notion that we, as ethical beings, have a responsibility to restore species (both plant and animal) because we recognize the value of the natural world.

However, all species live in a world wherein the dominant land use convention is economic. This Lockean view of land as property—that its ultimate use must be to derive profit or convenience (Locke 1963)—still constitutes the dominant land use paradigm in America (see Sax, Chapter Ten, this volume). Should an endangered species threaten to subvert this paradigm, either the law or the gun will amend. Thus, if a reintroduced wolf appears to threaten an economic interest (a cow), that wolf, for right or wrong, can be shot legally. In this way, the act prioritizes economic over ecological and other values. This "law of the tooth" is a rural reality, and of course this is where reintroductions take place.

Although urban Americans increasingly define parameters for environmental policy, they reflect the general population's lack of understanding of even the most basic biology. National Public Radio reported the results of a survey conducted a few years ago at the National Zoo in Washington, D.C., that found that more than 80% of the respondents did not understand the purpose of pollination. Even in my own teaching, in a rural state, more than 90% of my students have never learned to identify the trees in their own backyards. In the East, trees are the dominant feature of the natural landscape. Without even the most rudimentary appreciation for the workings of the natural world, how can we expect to find informed opinion on ecological restoration by surveying an increasingly urban, uninformed society? How will people understand habitat loss when they know so little of habitat?

As a result, experts often are called in to decide an issue. They understand the laws and the biology, and they make an informed set of decisions about restoring ecosystems, for example, or helping endangered animals recover. Does this expert handling accomplish any long-term goal? Skepticism is called for here because so much of this decision making is done while peering through a narrow lens at an extremely complex system. The system integrates the natural and built environments, incorporates human and wild dimensions, and links together the strivings and the debris of all life. A narrow "endangered species restoration" lens potentially obscures background conditions that may foil the plan in the long run. The Florida panther recovery program is a good example.

There are fewer than 40 Florida panthers left in the wild, primarily in south Florida. To help these remaining panthers recover, we have spent millions of dollars to bring them back. The question is, back to what? Every year, south Florida welcomes another 100,000 or so new residents, each one of whom feels entitled to a piece of the subtropics. They demand golf courses, malls, and new housing as well as countless other conveniences, all of which result in waste, pollution, and habitat loss. Long-term restoration of the Florida panther must attend to these issues as well as zoning, population growth, and residue leaching from farmlands. The focal lens for restoration projects must be broadened to address the impacts of these issues on the target species. A brief examination of leachates from farmlands provides a case in point.

Biomagnification of pollutants and inbreeding depression[1] seem to be linked to cryptorchidism in the panther (Harrison et al. 1997; Litchfield 1993). Mercury from both runoff and incineration is taken up in aquatic plants, concentrated in the tissue of fish, and further concentrated in raccoons, which eat them and are themselves an important component of the panther diet. Because the panther resides at the top of the aquatic food chain, each meal can contain a toxic concentration of pollutants in the tissues of its prey. This is caused by social and economic factors beyond the scope of the restoration lens. In addition to these problems, the number-one killer of panthers is cars; the number-one long-term limiting factor is habitat loss. The aggravating factor common to both is human population growth. The benefits of restoring a top predator don't necessarily trickle down through the system, particularly if the system's future is doomed to be compromised by social and political factors beyond the current purview of wildlife agencies. Perhaps a more holistic approach will involve wildlife experts working with family planning agencies.

Florida, like many other rapidly developing states, has some wildland left. But these wildlands are increasingly islands in an ocean of sprawl. We regard these remaining open spaces as the appropriate sites for our restoration projects and reintroductions, but relying on these spaces may, once again, defeat our long-term goals.

Thoreau said, "In wildness is the preservation of the world" (Thoreau 1990, 187). Following this thought, we have set aside our magnificent wild landscapes hoping to preserve nature to safeguard it from human encroachment. In these preserves, human settlement is forbidden, as is human interference in natural processes. These lands become sanctuaries for wilderness, they shelter species on the edge, and they are set aside as monuments of pristine nature that are untouched by or healing from the influence of human civilization. By restoring wolves (say, into our parks and refuges), we hope to redress old ecological wounds and return balance to the natural world. The message is clear: Nature is pure and healthy when secured from human intrusion. It is into these lands and with this ideology that large carnivore reintroductions and recovery programs are proposed (see Dizard, Chapter Eight, this volume).

In a recent essay, William Cronon (1995a) suggests that by separating humans and nature, by setting aside wild nature and investing those wilderness areas with sacred value, human participation in those systems renders the "sacred" profane. We reintroduce wolves into parks or refuges with the idea that they will be safe there and accord protection to them by the very fact that they are in a park, as long as they stay in the park. Should they venture beyond the boundaries, there are no guarantees.

This sets up a dichotomy between humans and wild nature that drives a wedge into the heart of what we are trying to accomplish. We would like to see wolves and humans coexist. But we place wolves apart from humans and the message is clear: That's the only way it's going to work. You can't place wolves where humans are; people would kill them. Parks, then, become refuges for wild nature; implicit here is that because humans are no longer part of that nature, we have no place in the "nature" occupied by wolves. Likewise, on our side of the boundary marker, wolves are not tolerated. When wild nature frees itself from arbitrary lines on a map and crosses over to our "nature," our old fears percolate up—through stockyards and pasture, through woodlot, corn-field, and meadow. By introducing wolves back into a park or a refuge, we communicate an idea about nature that sets us, once again, apart from nature and thereby sets wolves apart from us.

Proponents of natural recovery have the opposite goal. We want to be able to live *with* wolves. If wolves return on their own to a place, it is because the place has what they need. Wolves need a sufficient prey base and the use of the habitat their prey need. Given a chance to live, wolves have proven to be adaptable creatures. By affording a recovering species the chance to live, by tol-erating a natural recovery, we have shown that we can move over and share our privilege with another carnivore. We showed it in Minnesota, Wisconsin, and Michigan as wolves made their own return, and we may show it someday in New England as wolves try to come back here. There is something respectable about this. There is something to be said for "honoring nonhuman nature as a world we did not create, a world with its own independent reasons for being as it is" (Cronon 1995a, 87). This is an antidote to our arrogance in thinking we can and should control the patterns of nature. Wolves will return if they should be there, and people will be more likely to accept them that way. They have said so.[2]

A closer look at natural recovery in the Great Lakes states will help us con-sider the question in the Northeast.

Natural Recovery in the Great Lakes States

By the time the ESA was passed in 1973, the gray wolf, *Canis lupus,* had been extirpated from the Lower 48 states, with the exception of a remote corner of the Superior National Forest in Minnesota and Lake Superior's Isle Royale. Wilderness formed a refuge for this population, and it was there that perhaps 400 to 500 wolves still hunted deer and moose (Mech 1970). When pack size pressed individuals to leave their home territory, these dispersers pushed out from the wildlands, to be met by traps and bullets intended to confine wolves to their isolated domain.

With the passage of the ESA, however, the wolf was accorded full protection, requiring the federal government to develop recovery programs to ensure the wolf's return. Wolves became instantly glamorized as the emblem of endangered species protection, and with this protection wolves began reclaiming the North Woods of the Great Lakes states.

In the Upper Peninsula (UP) of Michigan,[3] 20 years had passed since the last breeding pair of wolves had been documented. The bounty on the wolf had been lifted in 1960 and the State of Michigan had granted protection to the wolf in 1965 (Hendrickson et al. 1995). Only the tracks of lone wolves, probably immigrating individuals from Canada or possibly survivors from the bounty days, were being reported (Thiel and Hammill 1988). Their tracks appeared on logging roads in scattered areas around the UP, but the news was interesting enough. It offered hope that perhaps a natural recovery was under way.

In the years after the passage of the ESA, the stage seemed set for a natural recovery to occur in the UP. In 1974, however, several researchers (Weise et al. 1975) attempted to translocate a small pack of wolves from Minnesota to a remote forested area in the central UP in the hopes of eventually establishing a viable population. Within 9 months, all four wolves were dead: Three had been shot and one had been killed by a car. The protection from which these wolves might have benefited under the ESA was thwarted by a vocal and angry minority who resented the intrusion of the government, which reintroduced wolves without soliciting their opinion. The bumper sticker "Save a Deer—Kill a Wolf" was intended to motivate hunters and alert government agencies.

Meanwhile in the woods, however, wolves were making a comeback on their own. The wolves in Minnesota had been increasing in number and expanding their range. Throughout the 1970s and 1980s, wolves slowly filtered southward through Minnesota's forests and backlands, picking their way across agricultural land and through the backyards of second homes. Many perished crossing the north–south highways of eastern Minnesota. When the survivors arrived at the south end of Lake Superior they located a narrow corridor into northern Wisconsin where several packs eventually established territories. As land in Wisconsin slowly became colonized by these wolves, new immigrants pushed through northern Wisconsin and began arriving in Michigan, attracted by its plentiful prey and forested habitat.

This remarkable process of natural recovery is underscored by the wolf's ability to disperse and adapt to new situations. In 1972 persecution still kept wolves bounded within the Superior National Forest, but once protected, the wolf expanded its range and within 20 years had reclaimed former territory in two adjacent states. Dispersal may be motivated by many factors, including social pressures within the pack or low availability of prey in an area, which can aggravate social pressures. Wolves are extremely vulnerable during dispersal because they are moving more frequently, passing through unknown territory, and their behavior can be less guarded and more erratic (Fuller 1989). Nevertheless, a determined wolf trying to locate a new territory can accomplish single straight-line journeys of 129 miles at a time. The longest movement of a gray wolf is an astonishing 620 miles (Ballard 1983).

In 1989, the discovery of what came to be known as the Nordic Pack signaled the return of wolves to the UP. Throughout the 1990s, breeding pairs established packs and continued to claim territories from the Wisconsin border eastward across the UP. From the few lone wolves whose tracks were reported in the 1970s, the wolf population on the UP has grown to include 216 animals as of late winter 1999–2000 (Hammill, personal communication, 2000).[4]

One of the important lessons learned from the wolf recolonization in the Great Lakes states is that its success can be attributed to the fact that the wolf came back on its own four feet. The only help the wolf received was the protection of the ESA. The State of Michigan has not offered any compensatory funding to farmers for livestock lost to wolf predation (which has not been much of a problem as of this writing). According to Hammill, the tolerance of deer hunters and landowners toward the wolves has made all the difference. Their support for wolf recovery arises as much from a strong public and hunter education program as from their appreciation for the animal making its own return; this constitutes a fundamental difference between reintroductions and natural recoveries. With political issues set aside, biological and social issues pertinent to wolf recovery can be resolved.

Wolf Restoration in the Northeast

The attention of wolf restorationists has turned to the Northeast, where it appears that a natural recovery of wolves from Canada into the United States is not likely because of significant habitat loss along the Canadian border (Harrison et al. 1997). In southern Ontario, much of the land is agricultural, and with no legal protection in Canada, most wolves are eliminated before they reach the U.S. border. Additionally, emigrating individuals from Canada into the United States would have to cross the St. Lawrence Seaway, a formidable barrier. Thus, a wolf reintroduction into Adirondack State Park in New York and the North Woods of New England is being considered. However, a confluence of social and biotic factors unique to the Northeast may militate against successful wolf reintroduction. A comparison to the Great Lakes wolf recovery program highlights the difficulties in the Northeast.

In the Great Lakes states, the gray wolf, *Canis lupus nubilis,* is large, averaging between 75 and 110 pounds. It coexists with the coyote, *Canis latrans,* which in the Midwest averages between 35 and 50 pounds. Both species prey on deer and beaver, with wolves also taking moose and coyote relying more heavily on rodents, birds, and vegetable matter. The wolf has exterminated the coyote in local areas throughout the Great Lakes, the size difference between these two species favoring the wolf.

In Adirondack Park, the most likely candidate to be translocated is the small Algonquin wolf, *Canis lupus lycaon* (which averages only 55 to 90 pounds) from Algonquin Provincial Park in Ontario. According to Dr. John Theberge (Theberge and Theberge 1998), deer are a key component in the diet of wolves in Ontario, and during the winter, when deer migrate out of Algonquin Park, the wolves follow. Because this small wolf is more disposed to hunt deer, it is assumed that the deer in Adirondack Park would be its main prey base. However, when Adirondack State Park was created in 1885, unregulated logging had laid waste to much of the timber resource. Following the creation of the park, the New York Legislature declared a large portion of Adirondack Park to be "forever wild," which placed the publicly held timber resources in the center of the park off limits in perpetuity. As a result, this land (approximately 60% of the park) has returned to more mature growth; the privately held land may still be logged, developed, and used for agriculture. Because older growth typically does not support many deer and because the current proposals would release wolves into the interior of the park, where the older growth is, it appears that the wolves, searching for prey, would hunt where the deer are, which is also where the people and the farms are located. A successful wolf reintroduction into Adirondack State Park would require a mandate in support of wolves by the private landowners in and around the park as well as a willingness on their part to share the ungulate resources. There are further difficulties for a wolf reintroduction into the northeast.

The current proposal would also reintroduce wolves into the North Woods of New Hampshire and Maine, where moose predominate. Deer, the main prey of the Algonquin wolf, rely on the forests and agricultural land in the southern portions of New England. According to the State of New Hampshire 1997 deer harvest statistics,[5] wildlife management units in the northern parts of the state record harvests a fraction of the size downstate. Algonquin wolves may not be big enough to effectively prey on North Woods moose and so, as in Ontario and Adirondack Park, they will go where the prey is and be drawn south into agricultural land to hunt deer.

Perhaps the most important point is that any reintroduced wolves would have to share an ecosystem currently saturated by the eastern coyote, *Canis latrans, var.* This coyote is larger than its western counterpart, averaging 40 to 60 pounds, with some individuals weighing more than 70 pounds. The eastern coyote may be the result of interbreeding with the Algonquin wolf. This may have occurred when the coyote migrated east after extermination of the wolf in New England at the turn of this century. This would explain their larger size, stronger jaws, and tendency to pack and hunt deer effectively. These traits also may be the result of a rapid evolution taking place in the coyote population in the Northeast, which is selecting for larger size and greater sociality than its western counterpart, enabling it to take larger prey.

It was expected that the wolf in the western and Great Lakes states would displace the coyote, and this has occurred in Yellowstone (Smith, personal communication, 1999)[6] and in local areas in the Great Lakes (Hammill, personal communication, 1999). Some predict that the same scenario will result in Adirondack Park and in New England, but I am skeptical, considering the niche overlap between the eastern coyote and the Algonquin wolf, their similarity in size, and the density of the coyote population in the East.[7]

Some scientists and conservation groups have assumed that *Canis lupus nubilis,* the wolf occupying the Great Lakes states, would be the appropriate subspecies to reintroduce back into the Northeast. However, recent genetic studies in Canada (Theberge and Theberge 1998) suggest that the wolf in Algonquin Park may in fact be the ancestral wolf of the Northeast. Even more surprising is the news that this wolf may be a red wolf. If this is supported, the small Algonquin wolf will make big news in the wolf world.

The red wolf, *Canis rufus,* was found in the southeastern United States and historically ranged as far north as Pennsylvania (Nowak 1979). Persecution and habitat loss limited its numbers and range significantly so that in the 1970s, the red wolf was placed under the protection of the ESA. The taxonomic status of the red wolf has engendered much discussion because it was thought to be a hybrid of the gray wolf and the coyote (Wayne and Jenks 1991) and that a continued hybridization with coyotes threatened to engulf the remaining wild population. By 1970, the U.S. Fish and Wildlife Service (USFWS) captured the remnant wild population of red wolves living in the southeastern United States and, selecting only the purest specimens, began a captive breeding program and successfully reintroduced several pairs into Alligator River National Wildlife Refuge in North Carolina. The Alligator River wolves were considered the last wild population of red wolves on the continent.

The new Canadian genetic study may extend the original range of the red wolf northward to encompass the eastern seaboard and southern Ontario (Brad White, personal communication, 1999).[8] It also may resolve the persistent question of which wolf is the "right" wolf to restore into Adirondack State Park and the North Woods of New England. If the ancestral North Woods canid is *Canis rufus,* the red wolf, returning this small wolf into a landscape fully occupied by the large eastern coyote raises serious biological considerations. For example, if coyotes in the Northeast rely on deer as an important component of their diet, what will be the cumulative impact on the deer herd of two sympatric competitors sharing the same resource? Also, because these two predators are so close morphologically, what will prevent interspecific breeding between the two populations or genetic swamping of the small red wolf by the eastern coyote?

Perhaps the most important point is that the eastern coyote appears to be evolving rapidly to fill the niche of the wolf. We must remind ourselves that the eastern coyote hunts the fields and woods of the Northeast largely because of our actions of 100 years ago. We cleared the land of forests, wolves, and their prey, moose and elk. With mostly cleared land, the deer returned and their population soared. With wolves removed and with little competition for this resource besides the human hunter, a new, opportunistic predator, the coyote, migrated in to fill the void we created. The coyote has now evolved in size and behavior into an increasingly more wolflike coyote. How fortunate we are to be witnessing the evolution of a predator within one human life span. However, the wonder of this phenomenon seems to elude many biologists and policymakers.

As a part of nature, we eliminated one animal that nature has replaced with another. We value the wolf but dismiss the coyote. In most New England states, coyotes can be trapped, snared, and shot 365 days a year, and in New Hampshire they can be killed at night from January to March, during their breeding season. We persecute the coyote while we urge compassion for the wolf and refuse to recognize this discrepancy. Perhaps we dismiss the coyote because its success offends us. Certainly our attempts to reduce coyote numbers have resulted in defeat for wildlife agencies, whereas eliminating the wolf proved an easier task.

Conclusion

Because we have changed our minds and values, valuing one species, the wolf, over another species, the coyote, we advocate for yet another manipulation that we feel would be more appropriate than the coyote for the Northeast. Why don't we spend our time and money on projects offering real long-term consequences such as developing binational agreements to purchase easements on land between Canada and the United States? This *would* provide migrating creatures corridors for movement, dispersal, and recovery. This may be a harder sell for conservation organizations, but if they truly educated their membership on the issues, people would support it. Teach the people to wonder at rather than disparage the resourceful predator howling in our woods; in learning to appreciate the coyote, they will extend tolerance, and perhaps even welcome, to the wolf, should a natural recovery be in the stars over the Northeast.

Endnotes

1. Inbreeding depression occurs when the number of individuals in a population decreases to the point at which matings occur between closely related individuals (between parents and their offspring or between siblings). The results of inbreeding depression matings are fewer offspring or young that are weak, experience greater juvenile mortality, or are sterile (Charlesworth and Charlesworth 1987; Ralls et al. 1988).

2. In fact, hunters in Michigan have been supportive of natural wolf recovery, claiming that sharing the woods with wolves has heightened their hunting experience. Public attitude surveys have indicated that the people of Michigan favor wolf recovery as well, citing specifically their support for a natural recovery rather than a reintroduction program (Kellert and HBRS 1990, 1991; Schadler 1994).

3. The Upper Peninsula (UP) of Michigan is 16,500 square miles of forest land, bog, and scattered human settlement. Like Wisconsin and Minnesota (and the North Woods of New England), the UP is in the Canadian biotic province, dominated by northern hardwoods with a spruce and fir subcomplex. Timber harvesting provides the foundation for many of the local economies and creates prime habitat for deer and moose, the prey of the wolf in the Upper Great Lakes.

4. According to Jim Hammill, the wildlife biologist heading the wolf recovery program on the UP, the recovery has had some setbacks. During the winters of 1995–1996 and 1996–1997, the deer herd suffered significant losses because of malnutrition. The wolves whose territories experienced the decline in deer correspondingly died of starvation, complicated by the onset of mange. Although the deer made a slight comeback during the spring of 1998, hunters may have grown alarmed over their declining numbers and, perhaps in response, killed three wolves. The reason for these deaths still is not certain, but the killings in any case seem unwarranted. Wolves take only a small percentage of the deer on the UP, averaging 18 to 20 deer per wolf each year for a total of 2,800 deer per year. Compare this to the 41,000 deer killed by hunters during the 1998 harvest season and the 15,000 deer killed by cars in the same year.

5. Information provided by Steve Webber, the deer biologist for the state of New Hampshire.

6. Doug Smith is the director of the Wolf Project in Yellowstone National Park.

7. According to Steve Webber, thousands of coyotes are living in New England, but no studies have been done to determine their numbers or their impact on other wildlife populations.

8. According to Brad White, the geneticist from McMaster University in Ontario who works with John and Mary Theberge on their wolf research in Algonquin Provincial Park, a gradient of wolflike coyote to coyotelike wolf might exist

between New England and Algonquin Park. The wolves residing on the outskirts of Algonquin Park have enough coyote genetic material that they could threaten the genetic integrity of the wolves within the park, which may be the most northerly population of red wolves on this continent, existing as an island population in a sea of coyote genes.

Part V

Wildness: The Wolf, the Human, and the Meaning of Responsibility

The three chapters in Part V examine the threat and promise of cultural beliefs about wolves and other wild things.

In Chapter Fourteen, philosopher Mary Midgley reflects on "The Problem of Living with Wildness" and the power and persistence of myths about wolves in human culture. Myths, she says, are used to identify or distance humans from the things that they revere or despise. Myths about the evil of wolves serve important symbolic and practical functions. When we cast wolves as the embodiment of human vice, that is, when we see wolves as evil, cruel, or vicious, we are given license to persecute, punish, and kill them. In so doing, we eliminate the wildness we fear in ourselves. The idea of wildness sums up all the antisocial tendencies that frighten us so much in ourselves, tendencies that are a constant threat to civilized life.

Likewise, she says, there is a countervailing tendency to honor and celebrate certain creatures as symbols of human glories and virtues. Lions are the prime example, constantly cited as showing noble and kingly virtues such as magnanimity. In the natural hierarchy, these impressive chosen species were held to reign as kings over their own particular grouping. Thus they represented order, not the disorder that was associated with other wild creatures.

Midgley emphasizes the symbolic role of wild animals to show how unreal and fantasy-laden are our inherited traditional ideas about them. What we need, she argues, are education programs designed to help us overcome not only our deeply symbolic fears of wildness but also the institutionalization of those fears in policies based on the assumption that civilization is essentially a justified war against nature.

Lately, she says, with the loss of biodiversity and other significant negative environmental impacts, it has begun to seem that we might in some monstrous sense win our bizarre war, that we might "defeat nature," cutting off the branch that we have been sitting on. To grasp this change calls for an unparalleled upheaval in our moral consciousness.

It will be necessary to replace the strong element of fantasy and projection that still influences our attitude to beasts such as wolves with a more realistic understanding of their situation and our own. One important step in this transformed consciousness, she says, will be to acknowledge a duty to nature or at least to future generations of humans.

Chapter Fifteen, "Leopold's Wildness: Can Humans and Wolves Be at Home in the Adirondacks?" is by philosopher Strachan Donnelley, director of the Humans and Nature Program at the Hastings Center. Donnelley draws on Aldo Leopold's concept of wildness to argue for a conception of ecological citizenship and, therefore, a more expansive concept of home for both humans and animals.

Echoing Midgley's call for a transformed understanding of obligation, Donnelley observes that wolf reintroduction is but a part of a much larger, challenging, and daunting issue: our long-term responsibilities to the region's human communities, natural landscapes, and ecological systems. How practically and morally are we going to bring humans and nature together in a viable and ongoing regional future?

Donnelley notes that this is not the first time wolves have figured in such questions and explorations. The death of a wolf in Aldo Leopold's essay "Thinking Like a Mountain" reoriented Aldo Leopold to an evolutionary, ecological, and biotic world view that includes human individuals and communities and fosters long-term responsibilities to the integrity, stability, and beauty of the land. Leopold's land ethic is an ecological, philosophical, and ethical proposal to be explored, not a dogma or ideology summarily to be accepted or rejected. What does it say morally about humans and wolves living together in nature? How adequate are Leopold's questions and answers to the reintroduction of wolves into the Northeast?

As a small step to addressing these questions, Donnelley examines Leopold's conception of wildness. In their historical interactions, wild things, alongside the abiotic elements of the natural world, have been world creators, engendering a specific home environment (ecosystems, bioregions, the land) for present and future evolutionary and ecological life. Humans are only late and problematic actors on this worldly scene. To be in touch with wild things—whether wilderness areas, large predators, birds, or wild flora and fauna as such—is to be in touch with the historical and ongoing cosmogony and our worldly and human origins. Wildness here is characterized by and has the value or emotional valence of roles played in the cosmogonic, evolutionary–ecological processes, the historical engendering of the diverse forms and capacities of life. This is the wildness of mountains, wolves, and more. To be with wild things in their habitats is to be at home in a very particular, primordial, and spiritual sense.

This world view radically transforms our usual sense of home, place, and situation, says Donnelley. We ordinarily understand our home to be in the here and now. In a certain sense this is true. Our immediate life is locally embedded in the here and now, in this particular place, time, and region. But Leopold would have us expand our spatial and temporal horizons. Both human and natural histories have generated or built our present selves, communities, and environs, and we are in the process of building the indefinite future. Thus our home, place, and situation cannot really be just here and now. We live in a much larger spatiotemporal, natural, and cultural edifice, with many rooms and passageways to the past, present, and future.

This change of perspective on ourselves and our world is not philosophically and morally innocent. Rather, it is a radically transforming sea change in viewing things, involving numerous "gestalt shifts" in valuation. From the perspective of the becoming, evolving land (our true, final, and primordial home) wolves and other large predators are seen as good (good for the ongoing biota), not bad, and deer and cattle are morally problematic. The former help uphold the ecological well-being of the land and the rich, reverential, respectful, or fearful liveliness of humans. The latter degrade the land by overgrazing plant resources.

Moreover, our technological triumphs and our economically good lives (for some of us) become seriously suspect. They tend to make us forget our origins, neglect our worldly home, and threaten the biotic future of the land community. Herein lies the human and ethical importance of Leopold's world view and moral landscape.

Chapter Sixteen is by philosopher Ernest Partridge, research associate in the Department of Philosophy, University of California. In "The Tonic of Wildness," Partridge maintains that the question of wolf reintroduction cannot be confined to the potential economic impacts of the decision. Instead, any legitimate debate must take into consideration noneconomic values, which, he says, can be neatly summarized by a single word: *wildness.*

Drawing on both the aesthetic and the evolutionary contribution of wildness to our own human nature, Partridge argues that by enhancing wildness where it exists and reintroducing wildness where it is absent, we enrich our personal lives, our communities, and our culture, now and far into the future. In addition, he maintains that preserving wildness enhances both our capacity and our worthiness, as a species and a civilization, to survive on the earth.

Finally, Partridge urges us to recognize that we humans are both natural and unique in our possession of the capacities that define our moral agency. The failure to acknowledge and incorporate both our naturalness and our agency into a system of ethics is a drastic mistake. By "denaturing" ourselves, we give license to those who would objectify and thus exploit the nature "out there." On the other hand, by depersonalizing ourselves, he says, we divest ourselves of moral responsibility for nature by regarding ourselves as objects captivated by and helpless in the stream of natural cause and effect.

Wildness, which once contained our species, now is contained by our civilization and survives only through our sufferance. Underscoring one of the major themes of this section, Partridge makes it clear that our moral responsibility must extend beyond humans to include the rest of our natural world.

Chapter Fourteen

The Problem of Living with Wildness

Mary Midgley

This chapter is largely about the symbolism connected with any project to reintroduce wolves. I shall not try to say anything here about the practicalities of such a decision because these must be left to people who have studied and debated the question. But I do want to say something at the outset about the interesting question of priorities.

How does this project rank compared with other possible ways of saving the environment? Is it more or less urgent than, for instance, finding a way to reduce pollution or save fossil fuel? What about the general need to save habitat? Should we always concentrate on preventing damage before we try to reverse the injuries we have already done? This question about priorities surely is a real one, something we ought to consider whenever we are choosing which causes to promote. But I think the answer to it is not quite as simple as it may seem. In the first place, there must always be some division of labor. There are always many good causes with roughly equal importance. In the second, these various causes often work to help one another. In particular, environmental projects that are psychologically gripping can bring home the importance of wider issues to people who otherwise would not take them in. Thus, when Rachel Carson's book *Silent Spring* (1962) made clear to a wide audience that insecticides were destroying songbirds, the extent of the general danger reached them more sharply than it would have if they had merely been told generally that the soil was being degraded. Even though that general degradation included a practical threat to their own welfare, they would not have seen that threat so clearly without the dramatic and unexpected reference to the birds. In this way the need to save songbirds pointed up ideals that supplied motivation for other, less exciting projects. Drama is not necessarily a distraction from practicality but can sometimes promote it.

Problems of Living with Otherness

I think the ability of drama to highlight practical concerns may be particularly true in the case of wolves because it is one on which our motivation has been exceptionally confused, not by the absence of drama but by a preexisting drama that is almost entirely misleading. We therefore need the new drama to correct the old one. In general, we have always been ambivalent about the natural world, particularly about its other large inhabitants. On one hand, we know that we are part of that natural world, that we owe our lives to it, and that it continually pours out treasures that delight us. On the other hand, it is also a potent source of death and danger. Being physically weak, our ancestors had good practical reasons to be alarmed about many of the creatures around them. But beyond any practical threat they posed, wild creatures have always been seen as powerful symbols, vessels filled with disturbing meaning. The strongest of them, those that are most often depicted in cave paintings, obviously had a particular grip on the human imagination. But even smaller and less alarming creatures could be seen as a psychic threat simply because they represent a kind of life so different from our own. They are mysterious, and mystery can always mean danger.

In very early times people seem to have dealt with this threat, as surviving hunter–gatherers still do, on the principle that if you can't beat 'em you should join 'em. They identified with these potent and mysterious creatures, propitiating them through rituals and trying to tap their peculiar force by means of sympathetic magic. Thus they tried to domesticate the alien power, to make it seem less alarmingly alien and external. Totemism is a systematic attempt to defuse the psychic dangers presented by otherness in this way. Later, however, as people developed their peculiarly human skills and set up larger communities, this kind of identification seems to have become harder. People whose way of life has become quite different from a porcupine's eventually find it hard to think of the porcupine as a close relation (Midgley 1978).

At this point the stark fact of otherness emerges more strongly. The human way of life begins to be seen as seriously different from that of other creatures. That change probably becomes more marked when people start keeping flocks and herds. When you depend on the produce of your domesticated animals, you can no longer afford to identify with other animals who might threaten your flocks, whether by attacking them or by competing for their fodder. And if you have sown crops, you want above all to stop those crops from being eaten by other animals.

This seems to be the point where the clash of interests between humans and other creatures became too sharp to be smoothed over by mythical identification. Of course the clash of interests itself goes back much further. Hunters have devised very interesting rituals to deal with it. For instance, Native American peoples who hunted buffalo and depended on buffalo products often maintained elaborate rituals that showed the buffalo accepting their fate in return for spiritual transactions that honored them and celebrated their relationship with their human friends. Whether or not the buffalo would actually have agreed to these deals, it is clear that such ceremonies served an important purpose for the hunters. Similarly, trees often are thanked, honored, and placated before being cut down.

This kind of symbolism has some practical importance in controlling human exploitation. But it also has a deeper psychological importance that I think has been less noticed. It shows that even people who regularly consume trees or buffalo, quite as a matter of course, already are feeling some guilt, some uneasiness about their systematic exploitation of these impressive beings. They sense that they do not fully understand their nature, that there is something sacred about them, and that there may be some danger, whether practical or spiritual, in simply subjugating them to our needs and wishes. These people regularly take life, but they are not wholly happy about taking it. There are myths everywhere about disasters following such exploitation. Shooting the sacred stag of Diana tends to have fatal consequences, and it is no good pleading that one didn't know that this particular stag was hers.

Nature Red in Tooth and Claw, Without and Within

Now this kind of guilt and uneasiness is well known. The interesting question that I want to ask is, What happens to it after people turn to pastoralism and agriculture? Does guilt of this kind just evaporate, or does it take other forms? It is worthwhile looking for these other forms because powerful motives like this usually do not vanish without trace. And there surely is such a residue to be found in the tendency of more settled people to morally downgrade the creatures they are freely destroying or exploiting.

As far as domesticated animals are concerned, this downgrading chiefly takes the form of contempt, a contempt that is sometimes mild and kindly but can sometimes be brutal. This contempt is expressed when one human being calls another a dog, cow, pig, goat, or sheep. But this contempt for wild animals can be far more serious. If these animals impinge on human life at all, they tend to be viewed not just as a practical nuisance but as embodying human vices. To speak of people as wolves, rats, vipers, sharks, or vultures is not just to say that they are troublesome. It is to accuse them directly of vice. And among these vice-denoting animals the most vice-denoting of all in our tradition has been the wolf, as one can check by looking up the entries under *wolf* in any quotation dictionary.

There are some interesting exceptions to this equation of wildness with evil. A few wild animals are given a favorable meaning. They show up an interesting ambivalence that we will look at later. But the equation with evil generally is so strong that it deserves attention. What is happening here is that nondomestic animals are being seen simply as embodiments of human vices. This projection absolves us of any guilt for killing them or for persecuting them when they are alive because they deserve it for their wickedness. This attitude still persists in ordinary discourse. In a popular book on wolves I once read that wolves trapped in medieval France were flayed alive with various appalling refinements. The author commented, "Perhaps this was rather cruel, but then the wolf is itself a cruel beast" (Midgley 1978, p. 27). Wolves are blamed for being sly and underhanded because they do not give their prey fair warning before they pounce on it. Similarly, rats are hated for being dirty, as if they had deliberately dirtied themselves out of malice before creeping in to infect the houses of their victims. In general, the animals are pictured as if they were human beings who had deliberately acted in such an antisocial way that they deserve to be killed.

This projection of human vices onto animals serves the purpose of making people in settled society feel justified in killing them for their own convenience. But I suspect that it also serves another, even more interesting psychological purpose. It provides settled people with a personification for the persistent vices in themselves that constantly make settled life so difficult. Killing the personification makes them feel that they have actually killed the vice. They are symbolically destroying their own wildness.

Projection and Self-Righteousness

The idea of wildness sums up all the antisocial tendencies that frighten us so much in ourselves—tendencies that are a constant threat to civilized life. This is surely why civilized people, even urban people, continue to be so keen on hunting, especially hunting predators. In such hunting, ritual and conventionalized forms surround the actual killing, disinfecting it from the dangerous social consequences that might otherwise follow from violence. It allows people to act out some of their own more savage wishes and at the same time to feel that they are destroying savagery in the outside world. The old photographs of big game hunters in Africa standing triumphantly with one foot up on various deceased creatures surely express this sense of having conquered something seriously noxious, and trophy heads on their walls convey the same impression, a fact that various cartoonists have happily exploited.

Just to show that I am not making this up, here is an example that shows this rather curious kind of self-righteousness. It comes from a journalist's account of a crocodile hunter called Craig who has been harrying a crocodile for many hours. He has fired several harpoons into it and is waiting for it to come up for breath and be killed. Meanwhile, he soliloquizes about it:

> "He's got the morality of a laser-beam" said Craig as we sat there, "the croc emerging from the egg will snap at anything that moves, no matter if it's a leech or a human leg." As he spoke he was tug-

ging on a harpoon line, trying to coax the beast below to move. "He's a dedicated killing machine, the killer of any fish, animal or bird." (Midgley 1978, 33)

This rather odd kind of moral judgment is not confined to hunters, of course. As civilization has expanded, as more and more land has been settled, the conquest of the wild been very widely seen as symbolizing the victory of good over bad, order over chaos, virtue over vice. Thus, in Tennyson's poem on the "Northern Farmer, Old Style" (Tennyson 1861/1987), the old man meditates on his deathbed about his life's achievements, and the only one that really satisfies him is that he has "stubbed up Thurnaby Waste." Skipping most of the dialect, here is his reflection on it:

Do but look at the waste, there weren't not feed for a cow

Nowt at all but bracken and fuzz, an look at it now—

Warn't worth nowt a hacre, an now there's lots of feed

Fourscore yows upon it, an some on it down to seed.

He did not just see wildness as unprofitable. He was not a mercenary man—indeed, the point of the poem is to contrast him with his mercenary son—and his mind was not on profit. He saw wildness as essentially alien and dangerous, and here Tennyson clearly applauds him. He was not alone in doing so. In the early twentieth century highly respected sages were still expressing this view—still busily grubbing up Thurnaby Waste—still speaking of our need to wage a general war against nature. Thus Freud wrote that the proper human ideal is that of "combining with the rest of the human community and *taking up the attack on nature,* thus forcing it to obey human will, under the guidance of science" (Freud 1930, §11). Marx had taken a similar line, and William James had made the same proposal in his famous essay on "The Moral Equivalent of War" (James 1911), where he said that the cure for human militarism was simply to redirect our aggression into the war against nature.

Of course, both James and Freud had a serious reason for making this suggestion. They both wanted to provide an outlet for human aggression other than fighting other humans, and they both saw how fearfully hard it was to do this. But providing a war against nature as a substitute takes it for granted that "nature" is something sufficiently like a human opponent to make this psychological shift workable. The drama is assumed to be the same in a way that can work only if the wildness in nature is assimilated pretty closely to the wildness of a human enemy. This involves making natural beings personify human vices. These prophets aren't talking only about deflecting aggression by the exhausting effect of straight physical labor. That idea would not call for the mention of war. They are talking about redirecting it into a different kind of hostility. Now Thurnaby Waste itself really cannot be seen as a suitable mark for personal hostility. It seems to me that the sinister enemy envisaged in the war against nature must really be the enemy within, the savage motives in all of us that cause us constant alarm by defeating our efforts to lead a civilized life. That, I suggest, is the kind of nature red in tooth and claw that really frightens us. When we find ourselves demonizing some group in the outside world to provide ourselves with an external enemy, that interior direction is really the direction in which we need to be looking.

The Other Side: Nature Benign

On the other side there is another symbolism, a countervailing tendency to honor and celebrate certain creatures as symbols of human glories and virtues. Some very strong animals have had a remarkably good press as images of human virtues, even though they were really dangerous. Lions are the prime example, constantly cited as showing noble and kingly virtues such as magnanimity. Thus Chaucer (1377/1899) wrote,

> For lo! the gentle kind (nature) of the lion!
> For when a fly offendeth him or biteth
> He with his tail away the fly smiteth
> All easily, for, of his genterie
> Him deigneth not to wreak him on a fly
> As doth a cur, or else another beast.

This kingly lion was often taken to stand for Christ, an idea of which C. S. Lewis's Aslan (Lewis 1950) is a recent example, standing for a special kind of wildness that is so grand as to be acceptable ("He's not a tame lion, you know"). Eagles, elephants, and the whale were also seen as symbols of kingship and thus as examples of kingly virtues. In the *scala naturae,* the natural hierarchy, these impressive chosen species were held to reign as kings over their own particular grouping (or, as we still say, their natural "kingdom"). Thus they represented order, not the disorder that was associated with other wild creatures. Bees got credit for their social virtues, being seen as civilized rather than wild. Doves, swans, and some other attractive birds were praised as examples of faithful monogamy. Horses and dogs sometimes were praised for their faithfulness, and this kind of praise was quite different from the language of symbolism because it was realistic. But it did not inhibit the far more common rhetoric of contempt. Thus Falstaff says, "I tell thee what, Hal, if I tell thee a lie, spit in my face, call me horse" (Shakespeare, *Henry IV, Part I,* act 2, scene 4, line 1214).

Why Symbols Matter

Why do symbols matter? I have labored this point about symbolism somewhat because I think it's important to grasp how unreal, how fantasy-laden our inherited traditional ideas about the wild animals of our planet have been until very lately and often still are. Throughout most of modern history, people in the West have divided these animals into a few simple groups. First there were the small creatures that make themselves more or less of a nuisance to civilized life, handily classed as vermin and calling for no detailed attention. Then there were certain grand and distant species who existed mainly as symbols but might occasionally be seen in the menageries of powerful humans, where they gained an extra symbolic meaning as indications of their owners' power. Neither of these groups seemed to raise any particular moral problems.

Then there were animals that were hunted, notably bears, wolves and other predators, wild boar, deer, and the exotic big game typically found in Africa. These creatures often were seen as deserving some kind of respect because hunters who took the trouble to study them often were impressed by their courage, intelligence, and other good qualities, yet they were still so wild that they could properly be killed. This sometimes led these hunters into a kind of ambivalence in which, without giving up hunting, they genuinely honored their game animals and took trouble to preserve their habitat, not just to do more hunting but for the sake of the creatures themselves. Theodore Roosevelt was a notable example of this dual role, but there have been plenty of other effective hunter–conservationists, a fact that conservationists who hate hunting need to remember.

Among these creatures, wolves retained a peculiarly sinister symbolism. Except for some Norse myths, sympathetic representations of them in literature and mythology are very rare in our tradition. The contrast with the lion is striking and probably results largely from the simple difference in their posture. Lions live and hunt on open plains, so they rely largely on their sight and often need to gaze attentively into the distance. This gives them that dignified, highbrow appearance that so much impresses human observers. Wolves, by contrast, are mainly woodland creatures, hunting largely by smell, so they work under cover much more than lions and do not have much occasion to raise their heads, hence the appearance of slinking, which humans found creepy. On top of this, Europe in the Middle Ages was largely lion free, whereas wolves survived and attacked flocks in some places until the nineteenth century. Although they rarely attacked humans, this caused them to be feared and it determined their special symbolic value. They stood for nature in that nature was opposed to civilization. And for a long time nobody seriously questioned that civilization was essentially a good thing.

Dogmatic Slumbers of Anthropocentrism

This world view was not really shaken in the Western tradition until our own time, even though protests against it have been rumbling for the last three centuries. It is impossible to exaggerate the enormous moral confidence with which Europeans invaded non-European countries, profoundly certain of their civilizing mission. Wordsworth and other Romantic poets did indeed protest (following Rousseau) that modern civilized life was diverging too far from nature, but they were widely seen as a dissident, impractical minority. During the nineteenth century a rather wider range of people did begin saying unkind things about civilization. Thus, Carlyle spoke sardonically of "the three great elements of modern civilization—gunpowder, printing and the Protestant religion" (Bartlett 1968, 575 a–b). Such comments have continued and sharpened in our own century. Will Rogers observed "You can't say civilization don't advance, however, for in every war they kill you in a new way" (Rogers 1949/1993). And a journalist once asked Gandhi, "Mr. Gandhi, what do you think of modern civilization?" Gandhi replied, "That would be a good idea" (Schumacher 1979, 2). But the received opinion among most people in positions of power was still the one expressed by Calvin Coolidge: "Civilization and profits go hand in hand" (Coolidge 1920/1993).

Only very lately has it begun to look as if this might not always be so. Only lately has it begun to seem that modern people might in some monstrous sense win their bizarre war, that they might "defeat nature," thus cutting off the branch that they have been sitting on, upsetting not only the poets but the profit margin as well. To grasp this change calls for an unparalleled upheaval in our moral consciousness. Even those of us who campaign to promote this realization do not really take it in completely.

During the last 50 years, starting from the first atomic explosions, a number of physical facts have gradually brought the change home to public consciousness. Most people who follow current events at all grasp, in theory, that there is a danger, and they want to do something about it. But we don't have the concepts ready to express this need. The change that is called for in our attitude toward nature is extraordinarily large—larger, probably, than any such moral change since the rise of agriculture. And our current ideas on that subject happen to be ones that have been deliberately narrowed, during the last four centuries, so as to make that change particularly difficult. Especially in political thinking, these ideas have been carefully framed to fix our attention solely on relationships between humans within society.

During the Enlightenment, Western thought largely broke with the idea that political obligation took its force from above, from our duty to God. Instead it derived that obligation from the social contract. It abolished the divine right of kings and substituted rational agreement between consenting citizens as the only basis of public duty. All civic rights were then supposed to be derived from that contract so that citizens had a duty only to fellow citizens. As far as human political life goes, this was surely a huge gain. The trouble is that it left out the rest of nature. Nature does not sign the contract, nor do its constituent parts. The rainforest and the Antarctic are not citizens at all.

Accordingly, Enlightenment political language makes it almost impossible to say that we can have any duties to them. Yet every sane person who looks at our current situation can see that we need to do something about them. The question is, In what terms are we to express that need?

This is the gap in our current moral thinking that leaves us mulling over problems such as the restoration of a top predator. To some extent we can bridge it by talk of self-interest, by thinking of these needs as matters of prudence rather than duty. But this bridge is too thin to carry all the heavy traffic we need. It is perfectly true that we are all in danger of disaster if we do not attack pollution, clean up the seas, and try to mend the ozone hole. But this looming disaster still seems to most people somewhat distant and diffused compared with, say, the danger of losing one's job, doing without a car, or otherwise becoming poorer by making a protest. There is a natural tendency to balance the personal risks and to hope that the world will last through our time.

It is possible to enrich contract thinking somewhat so as to widen our notion of prudence a little. For instance, we can say that we are all parts of a society that includes posterity, so that our own interest includes that of our descendants. The idea of signing an agreement with possible future people is not a very clear one, yet perhaps the contract often is conceived in that more generous way. But if that is indeed the intention, then we are no longer talking in terms of hard-nosed self-interest. This kind of contract thinking has already become a good deal less reductive, less crude, and less distinct from morals than it sometimes tries to appear.

In that more idealistic spirit, people have sometimes enlarged it further to include talk of a universal contract with nature, a basic agreement we all make by accepting the gift of life. But at this point the myth of contract becomes somewhat thin and visibly mythical. It is important in any case to remember that this talk of contract always *is* just a myth, and it does not seem to be quite the myth we really need here. Its language is not direct enough for the duty we feel. When we see the need to save the whales, the redwoods, or the Great Lakes, it is these things that we have to try and save. We are not trying to fulfill a contract with a remote and abstract nature. The claim on us is felt as quite immediate.

The Religious Background

Most cultures other than ours have acknowledged such claims and have dealt with them under the heading of religion, or at least something in the wide category that anthropologists call religion. However, their beliefs about them do not necessarily commit any of the distortions that made Enlightenment thinkers so hostile to religion. They don't necessarily even involve reference to any personal deities. Buddhist and Japanese reverence for nature is not theistic at all; it is direct. (I think that this is one reason why many Western people are now so impressed with Buddhism.) And in cultures where deities are involved, they are often just expressions of the natural human tendency to personify rather than damaging forms of superstition. Our own tradition does include quite a powerful strand of respect for nature as God's creation, a strand that is strong in Judaism and Islam and is now being reemphasized by Christian theologians. But in the early days of Christianity the church fathers pushed this notion into the background because they were so anxious to destroy any trace of nature worship. In spite of St. Francis, it largely remained there during the Middle Ages. And since the Renaissance, humanism has largely dictated that Christian as well as atheistic thought should confine moral consideration entirely to human beings. In our own time Christians are extending it to nonhuman animals, but they get little help from their own tradition in doing so.[1]

Repeatedly, thinkers in our culture have attempted to break out of this straitjacket. I have already mentioned the campaign for direct reverence for nature that stems from Rousseau. That campaign had a special resonance in the United States because people who came to America often were already in protest against many aspects of civilization in their native countries. The idea of civilization as such did not have for them quite the same kind of sacredness that it tended to have in Europe. Moreover, in early times they necessarily lived a less protected, urban life than they had done at home, and some of them came because they were attracted by the thought of this simpler existence. Thus many agreed with Thoreau and the founders of the great National Parks in celebrating the wilderness.

On the other hand, many of them had come deliberately to conquer wildness, not to accept it, and they devoted their huge energies to urbanizing things as fast as possible. Yet they still retained a kind of awe of the forces they were fighting. The most common compromise between these two sentiments is that which is expressed by shooting large animals. Another such compromise is the excitement felt at the idea of a wild frontier with wild inhabitants behind it, a frontier that is always there but is always being pushed further back. This idea is both a romantic celebration of wild nature and a declaration of war on it. The power of this double dream can be seen in the remarkable shift by which it has now managed to survive the taming of the western frontier and has been transferred to outer space, and the Restaurant at the End of the Universe, where it is probably safe from the same kind of interruption by reality. Today, however, we need a more realistic response.

What Then Must We Do?

That is why we need to develop new ways of thinking and why we get so desperately muddled when we have to apply them in detail. In the last few decades we have learned a lot of new words: *ecology, ecosystem, biosphere, symbiosis, Gaia,* and the rest. These words are framed to express a cooperative rather than a competitive relationship with other life forms, a relationship that is crucial in the workings of both our own nature and the nature around us, which our culture, since the Enlightenment, has refused to take seriously. Moreover, this cooperative approach clashes strikingly with the competitive individualism that has lately been so prominent in our social and political life. Two such different outlooks cannot be reconciled quickly. They cannot quickly educate each other. It is hard to see how we can combine them in a way that uses the best insights of both. Yet we can see that they do actually need each other because it has become clear that neither alone is adequate. Competitive individualism already is in grave trouble because it has become so extreme as to be impractical, as is plain from recent spasms of lethal lurching in the money markets. It has appealed to a one-sided romantic exaltation of individual freedom that badly distorts public life, and it is proving not to deliver private happiness either—a point Robert Bellah made clear in his impressive book *Habits of the Heart* (1985). And ecological thinking, for its part, has to grapple with the realities of a competitively organized human society if it is to put its restoration projects into practice.

Muddles about ideals are thus piled on top of practical difficulties. This is what makes projects of the kind we are considering in this volume so irredeemably messy. Speaking with the utmost caution and humility as an imported observer, I get the impression that what is going on in debates about wolf restoration is very important but that its importance may center on the change of attitudes it involves rather than on the details of what happens on the ground.

What is crucial about the change of attitudes is the negative side. We profoundly need to get *rid* of something. We need to get rid of the notion that all natural things are valueless in themselves, merely pretty extras, expendable, and secondary to human purposes. That notion is so radically bad that we must ditch it somehow, even though we don't yet have a perfectly clear idea yet of the ideals we must put in its place. We have partial and scrappy notions of the ideals on which we must work further, as always happens when people make a necessary shift of priority among ideals. We can make this move only if we jettison the exploitive attitude that has governed us so far. It is a habit of mind that our society desperately needs to reverse, and it has been an extremely deep one, as the symbolism I mentioned earlier on illustrates. It is an attitude expressed in countless customs in our lives.

To dig out something so deep in our psyches, we do indeed need to reverse it explicitly in practice. The painful words "We Were Wrong" must be not only spoken but spelled out in action, and it must be action with a strong symbolism that bears on the offenses that have been central to our crimes. That is why it is right that the Pope apologizes to the Jews for the Church's anti-Semitism, even though that apology may seem absurdly inadequate and disproportionate to the evil done. Similarly, it is right that people who are proved to have been wrongly convicted of offenses are rehabilitated after their death, even though this can no longer help the victims themselves. Moves such as these are not just futile hand-wringing over the past. They are ways of committing us to changing direction for the present and the future. Similarly, when today's Americans first save existing habitat and then go beyond this by calling back its previous inhabitants, that is what they seem to me to be doing. They don't do it as an isolated and artificial move confined to the wilderness but as an earnest expression of a far wider campaign that must involve enormous changes in human life.

Endnote

1. I have discussed this difficulty in regard to animals in the first chapter of my *Animals and Why They Matter* (1984). On the relationship between Christian contributions to this difficulty and those coming from the Enlightenment, see John Passmore, *Man's Responsibility for Nature* (1974), and Stephen Clark, *Animals and Their Moral Standing* (1997).

Chapter Fifteen

Leopold's Wildness: Can Humans and Wolves Be at Home in the Adirondacks?

Strachan Donnelley

There are some who can live without wild things, and some who cannot. These essays are the delights and dilemmas of one who cannot.
—Aldo Leopold, *A Sand County Almanac*

The proposed reintroduction of wolves into the Adirondack Park and north-eastern United States is a part of a much larger, challenging, and daunting issue: our long-term responsibilities to the region's (and the world's) human communities, natural landscapes, and ecological systems. How, practically and morally, are we going to bring humans and nature together in a viable and ongoing regional and global future? Humans and nature too often and for too long have been kept apart in philosophical and scientific explorations and ethical and social policy deliberations. As a result, we are inadequately prepared for the crucial task before us. This is a situation we must remedy.

This is not the first time wolves have figured centrally in such explorations. The death of a wolf in Aldo Leopold's essay "Thinking Like a Mountain" in *A Sand County Almanac* (1949) reoriented Leopold (in symbol, if not in fact) to an evolutionary, ecological, and biotic world view that includes human communities and fosters long-term responsibilities to the integrity, stability, and beauty of the land or the biotic community as a whole, which decidedly includes us.

Leopold and *A Sand County Almanac* remain directly relevant to the task of fashioning an ethic for the human and natural future. Moreover, we are reminded of an important but often forgotten philosophical truth. The meaning and significance of *conservation, wolves, wildness,* and *substantive ethical respon-*

sibilities do not exist in a conceptual or intellectual vacuum. They are embedded in a wider and philosophically fundamental view of the world and our place in the broader scheme of things. If we honestly want to embrace and use—or constructively quarrel with—Leopold's land ethic, we must appreciatively and critically understand his fuller world view. This effort is eminently worthwhile in itself and offers the opportunity to consider seriously whether, from an ethical point of view, we ought to reintroduce wolves into the Adirondacks and northeastern United States. Although we should expect no easy, definitive answers, at least we should be able to see our serious moral options.

Leopoldian Reality

Let us start by considering more closely Leopold and the wolf in "Thinking Like a Mountain" (1949, 129). Leopold had been professionally involved in game management in the Southwest, eradicating predators (wolves, bears, cougars) for the sake of increasing deer populations for hunters, if not paving the way for cattle ranching. While on a mountain trip, Leopold and his companions came upon a wolf crossing a river to join her grown cubs. Following the dictates of "wise use" game management and the trigger itch of young hunters, the group shot at the wolves, killing the mother. As Leopold watched the "fierce green fire" dying in her eyes, he was personally and philosophically humbled. The mountain and the wolf knew something that Leopold did not, and what they knew shamed him. Large predators have an ultimate significance and central role to play in evolutionary, ecological, and geologic time and the ongoing well-being of ecosystemic nature and the good human life. Leopold had previously been thinking, feeling, and acting in a wrong frame of reference. He had been animated by an inadequate and faulty view of the world. He had not taken a long-range biotic, evolutionary, and ecological perspective and did not appreciate the roles that wolves and other large predators play in the overall health of specific ecosystems (keeping prey species at healthy and adequate levels, preventing the overcropping of plant resources, helping the internally complex ecosystemic whole maintain a dynamic stability or equilibrium). Henceforth, Leopold knew better.

What precisely did Leopold know, think, and feel? What were his new and transformed world view and moral landscape? To approach these questions, I want briefly to appeal to some venerable philosophical terms and inquiries: ontology, cosmology, and cosmogony. Briefly, ontology is the study of the fundamental character of reality or being (for example, whether reality is essentially permanent, unchanging, and eternal or always in a state of becoming, change, flux, and process or some intermingling of permanence and change). Cosmology is the study of the structure (permanent or achieved) of the universe or cosmos. On the other hand, cosmogony is the study of the genesis or coming-into-being of the (structured) universe or cosmos. Among many others, the Greeks, including Plato, Aristotle, and the Presocratics, had much to say on these issues.

These seemingly esoteric but fundamental explorations are relevant to us because they are all deeply value-laden, have moral import, and form the backbone of cultural and ethical world views. For example, the cosmogonic myths ("true stories") of traditional societies provided a fundamental and socially crucial cultural and moral orientation to members of their communities (Eliade 1985). The myths related where the world and humans came from, who the community members were, what their role and significance were in the wider scheme of things, and what their fundamental moral duties were to themselves and wider reality. The Book of Genesis, for example, relates the cosmogonic myths of the Judeo-Christian tradition. Many "enlightened" moderns believe that we are well beyond such primitive cultural and traditional beginnings. Others, including Leopold, think that in this belief in our independence from myth, we moderns are deluding ourselves. Human individuals and communities endemically need some form of cosmogonic myth, some basic philosophical, moral, and spiritual orientation. Leopold offers us a wild cosmogony to undergird and help explain his land ethic and our ultimate moral responsibilities.

Leopold is not so concerned with the universe or cosmos as a whole as he is with our biotic earth. Worldly life is the experiential foothold for his ontology, cosmology, and cosmogony, which are informed by his first-hand experiences of nature and the Darwinian sciences of evolutionary biology and ecology. Darwinian biology either explicitly holds or strongly implies the following tenets, which find their way into Leopold's philosophy and world view (Mayr 1991).

- That nature has no established cosmic design or designer and is not ruled by cosmic teleology (purpose). Nature, biotic and other, creates its own forms in passing.
- That all biotic (if not other) things are multicaused on multiple spatial and temporal scales. There is no strict causal determinism à la Newtonian science, that is, physical antecedents totally and straightforwardly determining physical or material consequents. Rather, complex historical contexts and contingencies rule the biological scene.
- That there are no "species types" ("dog," "cat," "human being"), as espoused by traditional philosophical and theological world views, but only particular and diverse individuals living, interacting, and reproducing in particular "species" populations, wider communities of organisms, ecosystems, bioregions, and the biosphere as a whole. In sum, particularity and diversity, as well as biological and behavioral habits shared in common, characterize the historic biotic realm.

- That human individuals, communities, and cultures arise within historical evolutionary and ecological contexts and processes, that is, in dynamic interaction with nature or the wider biotic community. Humans have not appeared on the scene from an aworldly elsewhere.

Given this brief philosophical and Darwinian journey, we are now prepared to understand what wolves, mountains, the biotic community, and the land mean for Leopold scientifically, philosophically, spiritually, and morally. We can best approach this understanding from the cosmogonic perspective and Leopold's own cosmogonic myth.

The biotic community (all interacting biological entities, all complex and intermingling food chains, energy circuits, and predator–prey relations) or the land (all biotic and abiotic entities that together support life on earth) result from historical and ongoing evolutionary, ecological, biotic processes. In brief, we and all animate things exist amid and are involved in cosmogonic processes: the genesis or coming-into-being of earthly biotic reality. Life and biotic communities (flora and fauna, interacting individuals, populations, ecosystems) have created, and continue to create, the land. We humans, immersed as we are in nature and biotic processes and communities, have become increasingly important actors in this ongoing cosmogony through our burgeoning uses of natural resources and population pressures.

Leopoldian Values, Philosophy, and Ethics

What does this historical, worldly cosmogony, which includes humans, mean for Leopold? Practically everything, especially concerning the humanly and naturally good life and our ultimate moral responsibilities. I first want to begin with the meaning and importance of "wild things" for Leopold: mountain ranges, large predators (wolves, bears, cougars), and wild flora and fauna. In their historical interactions, they (along with the abiotic elements) have been world creators, engendering a specific home environment (ecosystems, bioregions, the land) for present and future evolutionary and ecological life. Humans are only late and problematic actors on this worldly scene. To be in touch with wild things—whether wilderness areas, large predators, birds, or wild flora and fauna as such—is to be in touch with the historical and ongoing cosmogony and our worldly and human origins. Wildness here is characterized by and has the value or emotional valence of roles played in the cosmogonic, evolutionary–ecological processes, the historical engendering of the diverse forms and capacities of life. This is the wildness of mountains, wolves, bears, cougars, sandhill cranes, grebes, prairie grasses, and more. To be with wild things in their habitats is to be at home in a very particular, primordial, and spiritual sense.

The Darwinian–Leopoldian cosmogony radically transforms our usual sense of home, place, and situation. We ordinarily understand our home, place, or situation to be in the here and now ("simply located," as Alfred North Whitehead would put it). In a certain sense this is true. Our immediate life is locally embedded in the here and now, in this particular place, time, and region. But Leopold would have us expand our spatial and temporal horizons. Both human and natural histories have generated or built our present selves, communities, and environs—think of organisms' genomes (genetic material) and habitats—and we are in the process of building (creating the ground of) the indefinite future. Thus our home, place, and situation cannot really (ontologically) be just here and now. We live in a much larger spatiotemporal, natural, and cultural edifice, with many rooms and passageways to the past, present, and future. This way of looking at our home ground should not really surprise us. Our immediate personal lives are dominantly characterized and informed by past memories and future expectations. The world (the land or biotic community) merely mirrors in a nonsubjective or nonexperiential way what we subjectively experience. In all probability we are what we are (bodily subjects experiencing a past, present, and future) because the world is the way it is, that is, a historical becoming in which past, present, and future are integrally connected. Human (and animal) selves arise out of the world and its fundamental (ontological and cosmological) characteristics, not vice versa.

This change of perspective on ourselves and our world is not philosophically and morally innocent. Rather, it is a radically transforming sea change in viewing things, involving numerous "gestalt shifts" in valuation. From the perspective of the becoming, evolving land (our true, final, and primordial home), wolves and other large predators are seen as good (good for the ongoing biota), not bad, and deer and cattle are morally problematic. The former help uphold the ecological well-being of the land and the rich, reverential, respectful, or fearful liveliness of humans. The latter degrade the land by overgrazing plant resources. Sandhill cranes and grebes are considered the wildest and most significant of birds: the great trumpeters and conductors of the orchestra and chorus of the grand evolutionary–ecological symphony (Leopold 1949). They carry us back to an aboriginal and immemorial past, to evolutionary origins before we humans arrived on the scene. We are in their debt for vividly disclosing our home in nature among wild things. (We *are* actually at home in nature and the world, notwithstanding existentialists and philosophers of "higher realities.") Moreover, our technological triumphs and our economically good lives (for some of us) become seriously suspect. They tend to make us forget our origins, neglect our worldly home, and threaten the biotic future of the land community. Herein lies the human and ethical importance of Leopold's cosmogonic account, world view, and moral landscape.

Wolves in the Adirondacks: Some Ethical Considerations

In this volume, several ethical issues emerge with respect to the proposed reintroduction of wolves to the Adirondacks. These ethical issues involve humans, animals, and nature severally and in their interactions. What can we learn from the preceding Leopoldian reflections about our ethical options and responsibilities, especially to the human and natural future? I offer the following considerations.

The Evolutionary Ecology of the Adirondacks

As we have seen, large predators, wolves in particular, have played a significant role in the evolutionary and ecological well-being of biotic communities and the land (the long-term well-being of the mountain). Would this be the case for the future ecosystemic well-being of the Adirondack Park, given the presence of humans (including hunters), other predators (coyotes), and likely prey species (deer and beaver)? Would or could wolves be integral to the ongoing health of the "Adirondack Land"? If so, this would be an especially strong moral argument in favor of their reintroduction.

Wolves and Their Welfare

Leaving aside the impacts on coyotes and prey species, for good or ill, what might be the impact on the wolves themselves? Could they live reasonably natural large predator lives? What if their introduction were successful and their population flourished? We know that wolves are far-ranging, that they would not by themselves keep within the confines of the park. (Leopold himself recognized that wolves could not be confined to parks.) Presumably wandering wolves would have to be managed, with a certain percentage killed or culled. Even if we could morally tolerate this, could we manage wolves in a morally acceptable and humane way? What would the culling or killing of individual wolves do to the social structures of wolf packs? For this, we would have to ask wolf ethologists, assuming (perhaps erroneously) that they have the answers.

Humans and Their Land Home

The possible reintroduction of wolves into the Adirondack Park with regard to human values is exceedingly complex, as we would expect. Many values and responsibilities, some inevitably in conflict, are involved. There are important if prosaic questions of economic impact on human communities and the safety of livestock, domestic pets, and perhaps humans themselves. (No doubt there are both moral pluses and minuses here.) However, there are other philosophically interesting and practically important issues, particularly relating to our long-term responsibilities to human communities and nature, to our Leopoldian home and its future inhabitants, human and other. Would reintroducing wolves spur us to adopt a long-term land ethic, to see the Adirondacks as an important bioregion or ecosystemic complex worthy of protection, scientific study, and wild recreational opportunities? Would this fuel a wider and robust global "humans and nature" conservation ethic? If so, bring on the wolves.

Or would the reintroduction and consequent managing of wolves transform the Adirondacks culturally into a Disney-like theme park, experientially cutting humans off from their evolutionary and ecological context, their Leopoldian home and origins so crucial to their spiritual well-being, humanly good life, and sense of ultimate stewardship responsibilities? If this were the likely outcome, might it not be better to expand our horizons to consider our home region to include Canada and Ontario's Algonquin Provincial Park, where we could meet wolves on their (now) wilder home turf, with freedom to roam north unimpeded and unculled? (The land knows no political and national boundaries, so why should we, the "plain members and citizens" [Leopold 1949, 204] of the land?) In the long run, would our journeying north (in thought or reality) serve us better spiritually and ethically than returning wolves to their old Adirondack home, where they may no longer fit so well, with limited or diminished prospects for the future?

These are the human nature, biotic community, land ethics issues that we as a democratic community—emphatically including Adirondackers and other legitimately concerned citizens—must face squarely. Note that these issues do not entail merely personal, economic, or political preferences (what we simply like or dislike). They involve genuine ethical problems or dilemmas. They deal with what is good and bad, right or wrong, for a human and natural reality that is laced with significant values and worth and is vulnerable to harm, ignorant blundering, or willful neglect. Our philosophical world view here makes all the difference in determining our moral responsibilities. To ignore or turn our backs on these ethical realities and corresponding imperatives would be morally and politically irresponsible, certainly no way to treat our Adirondack and wider evolutionary, ecological, and biotic home, preparing for all its future inhabitants.

Chapter Sixteen

The Tonic of Wildness

Ernest Partridge

We need the tonic of wildness. . . . At the same time that we are in earnest to explore and learn all things, we require that all things be mysterious and unexplorable, that land and sea be infinitely wild, unsurveyed and unfathomed by us because unfathomable. We can never have enough of Nature. . . . We need to witness our own limits transgressed, and some life pasturing freely where we never wander.

—Henry David Thoreau (1854/1990, 187)

Opponents of wolf reintroduction often argue that according to the calculations of cost–benefit analysis, a favored method of policy analysis, the wolf has no place in the Adirondacks. They point out that it preys on a profitable game species, the deer, and thus would adversely affect the economy of a region that depends heavily on sport hunting. Moreover, although the danger of wild wolves to humans has been vastly exaggerated, their introduction into inhabited areas is not without risk. In short, if we were to confine the controversy to economic considerations, they might well settle the issue: no wolves in the Adirondacks. The case for wolf reintroduction, which I support, must therefore appeal to noneconomic values, which can be neatly summarized by a single word: *wildness*.[1]

The Adirondack region, once logged over, now inhabited, and surrounded by human settlements, is not a wilderness and can never be a wilderness on any time scale relevant to the concerns or planning of our generation or its near successors. But the region can be wild and becomes ever more wild as the time of its exploitation recedes into the past and natural processes are allowed to take over and dominate the landscape.

A region that is managed by game laws and private property owners is less wild. A region that acquires its qualities through the uninterrupted playing out of natural forces is more wild. A deer population that is kept healthy through careful monitoring by wardens and researchers, and consequently by fluctuating hunting seasons and quotas, is less wild than a population that is culled naturally by predators.[2]

Thus a decision to reintroduce wolves into the Adirondacks would be a deliberate decision to enhance the wildness of the park. Why would we want to do this?

In this chapter I defend the thesis that by enhancing wildness where it exists and reintroducing wildness where it is absent, we enrich our personal lives, our communities, and our culture, now and far into the future. Furthermore, I argue that preserving wildness enhances both our capacity and our worthiness, as a species and a civilization, to survive on the earth.

An Adirondack Park with wolves would be a wilder place, and that is why the wolves should be reintroduced.

The Experience of Wildness

To begin, the value of wildness might be understood through an analysis of the natural aesthetic: the features of the experience of the wild.

The word *beautiful* is used to describe both fine art and natural landscapes. But the beauties of art and nature are radically different. Natural beauty, unlike artistic beauty, is uncomposed, unframed, and inclusive. Each of these qualities has important implications for the value of wildness.

Composition

Art is the product of the artist, whereas nature is the product of natural forces, both abiotic (e.g., erosion and sedimentation) and biotic (e.g., evolution). The composition of an art object is a deliberate creative act of the artist, although the creative act often draws on unconscious sources that astonish both artist and spectator. Art therefore addresses our humanity as a communication from artist to audience, one person to another.

Encounters with nature address a more fundamental essence: our naturalness, the sources and sustenance of our biotic being, including the precultural neural apparatus of our senses and cognition. The sight of eroded slickrock in the Utah canyonlands, the sound of cascading water in the Hudson rapids of the Adirondacks, the fragrance of a rotting log in an eastern deciduous forest, the fleeting glimpse of the wild wolves, the sound of their howl, and the still evidence of their work on the forest floor—none of this was composed with the purpose of communicating from one person to another. All of it communicates, but communicates what?

Most of the remainder of this chapter addresses that question. But as a beginning, the uncomposed message of wild nature speaks to us of vastness, independence, permanence, and ontological priority. With no creative artist in the landscape, we encounter instead the results of forces, in the past and still at work, that preceded our personal and cultural existence, formed us and sustain us, and will long survive us. In the face of the timeless and infinite, we are thus reminded of our own finitude. We encounter in undisturbed wildness what Edmund Burke called the Sublime, at once terrifying, invigorating, and morally instructive.

> The passion caused by the great and sublime in nature, when those causes operate most powerfully, is astonishment; and astonishment is that state of the soul, in which all its motions are suspended, with some degree of horror. In this case the mind is so entirely filled with its object, that it cannot entertain any other, nor by consequence reason on that object which employs it. Hence arises the great power of the sublime, that, far from being produced by them, it anticipates our reasonings, and hurries us on by an irresistible force. Astonishment, as I have said, is the effect of the sublime in its highest degree; the inferior effects are admiration, reverence, and respect. (Burke 1761, Part II, Sec. I)

Framing

Artistic works are confined by boundaries. The dance and the sonata are temporally bounded: They have a well-defined beginning and an end. The painting is bounded by the edge that separates it from the gallery wall.

In contrast, wild nature is unbounded. Gazing through the desert skies of a moonless midnight, we find no borders around the mantle of stars. The eastern horizon, we well know, hides still more stars that will soon come into view and, in its fullness, a cosmic sea that surrounds our insignificant planet and its minor star. Similarly, we know that beyond the horizon that we view from the mountaintop there is still more. The uncomposed wild landscape before us is without frames, without well-defined borders that define it as a separable object. Likewise, as we hear the howl of a wolf and watch the alerted deer bound out of sight, we understand that this episode is a snapshot in time, emerging from a natural past without defined beginning and merging into future without defined end.

Thus, in the presence of unbounded wildness, we are drawn into an awareness of ever more time and space. And in contemplation of time past, our imagination is cast before the time of our personal origin, and of the origins of our culture and our species, to the timeless foundations of all process, of all being.

Inclusion

The spectator of a work of art stands apart from the art object, for it is a separable object, framed and unified by the creative act. In contrast, in the presence of unbounded wild nature, we are drawn into the landscape as it surrounds us.

This is especially true as one becomes physically engaged with the wild, as a kayaker negotiating the rapids of the Hudson river, a photographer stalking the wild wolf, a hiker along the Appalachian Trail, or a skier carving into the virgin Rocky Mountain powder. In such cases, the subject–object boundary is obscured as one becomes, Zen-like, his or her natural environment.[3]

Thus might an encounter with wild natural beauty—uncomposed, unframed, and inclusive—add to abstract knowledge, the vital personal dimension of appreciation. Knowledge is intellectual, and appreciation is the aesthetic, emotional, and moral supplement to the knowledge that leads to action. The heavy smoker *knows* that he or she is taking a risk; *appreciation* comes too late, with the diagnosis. We *know* that the tropical rainforests are disappearing at the rate of an acre a second, but do we *appreciate* it? The *National Geographic* tells us that the Grand Canyon is a mile deep and 200 miles long; appreciation comes from sitting on the south rim, feet dangling over a thousand feet of sheer drop, looking across to Bright Angel Canyon. Books tell us of the eras and periods of geologic history; appreciation comes as we leave the rim of that canyon and walk down through the strata of frozen time toward the Phantom Ranch in Granite Gorge.

Aldo Leopold's moment of appreciation came as he saw "a fierce green fire dying in [the] eyes" of the wolf that he and his companions had shot.

> I was young then, and full of trigger-itch; I thought that because fewer wolves meant more deer, that no wolves would mean hunters' paradise. But after seeing the green fire die, I sensed that neither the wolf nor the mountain agreed with such a view. . . . We all strive for safety, prosperity, comfort, long life, and dullness. . . . A measure of success in this is all well enough, and perhaps is a requisite to objective thinking, but too much safety seems to yield only danger in the long run. Perhaps this is behind Thoreau's dictum: In wildness is the salvation of the world. Perhaps this is the hidden meaning in the howl of the wolf, long known among mountains, but seldom perceived among men. (Leopold 1949, 130–133)

Wildness and Human Nature

Homo sapiens is both a natural and an artificial creature. We disregard either aspect of our nature at our peril.

The Natural Endowment

I submit that our naturalness is beyond reasonable dispute. We breathe natural air, we are nourished by natural food, we respond naturally to the rhythms of life, and eventually we give back to the earth the matter it gave to us.

Significantly, our natural endowment long predates our artifice. Ten mil-lennia ago, an instant in geologic time, *Homo sapiens* established the first per-manent agricultural villages and began domesticating plant and animal species, thus pushing wild nature back to the perimeter of the village, then further still, until today only scattered remnants remain. These ancestors, who for all time before owed their survival to their adaptation to wildness, carried essentially the same genome that each of us possesses today. Thus wild nature, not the artifice that surrounds us today, selected our genes. Accordingly, writes biolo-gist Hugh Iltis, "like the need for love, the need for [the] diversity and beauty [of nature] has a genetic basis." He continues,

> The best environment is one in which the human animal can have maximum contact with the type of natural environment in which it evolved and for which it is genetically programmed without sac-rificing the major advantages of civilization. . . . Every basic adap-tation of the human body, be it the ear, the eye, the brain, yes, even our psyche, demands for proper functioning access to an environ-ment similar, at least, to the one in which these structures evolved through natural selection over the past 100 million years. (Iltis 1967, 887)

More recently, eminent Harvard biologist Edward O. Wilson has given this theory the name *biophilia*. He writes,

> The brain evolved into its present form over a period of about two million years, from the time of *Homo habilis* to the late stone age of *Homo sapiens,* during which people existed in hunter–gatherer bands in intimate contact with the natural environment. Snakes mattered. The smell of water, the hum of a bee, the directional bend of a plant stalk mattered. The naturalist's trance was adaptive: the glimpse of one small animal hidden in the grass could make the difference between eating and going hungry in the evening. And a sweet sense of horror, the shivery fascination with monsters and creeping forms that so delights us today even in the sterile hearts of the cities, could see you through to the next morn-ing. . . . Although the evidence is far from all in, the brain appears to have kept its old capacities, its channeled quickness. We stay alert and alive in the vanished forests of the world. (Wilson 1984, 101)

Biophilia lends depth and credence to the notion that we are natural crea-tures, for it adds to the unquestioned biotic needs for human life the intrigu-ing notion that the wild nature that selected our genes is needed to satisfy genetically programmed emotional and psychological needs as well.

But the nature that effected this selection is a nature that is fast disappear-ing because of our carelessness and greed, so that we may find ourselves in a world to which we are ill adapted. Thus it may be a deadly error to treat nature solely as a mere resource for our use, for to do so is to commit the deadly sin of pride, the hubris of regarding our artificial needs as of more fundamental value than the nature that is continuous with our selves. Science tells us otherwise. We are nature, and nature is us; "the world is our body."

If the biophilia hypothesis is correct, the implications are portentous: Destroying the final remnants of wildness will cast us into an alien world, devoid of the landscape that selected us and thus, in a fundamental sense, *is* us. Because we are unaware of the price of that alienation, we should pause before we commit ourselves to its payment. In the meantime, we are best advised to preserve the wildness that remains and to nurture its return to regions, such as the Adirondacks, from which it has been diminished.

Artifice and Agency

We are also artificial, through and through, not only because of the artifacts that surround us but more fundamentally because of the most basic human institution, language, which is the foundation of our mode of perception and thought, our funded knowledge, and moral responsibility.

Mark Twain once said, "Man is the only animal that blushes—or needs to." That observation was more wisdom than wit. Blushing is a response to the moral sentiments of shame and guilt. These sentiments, along with the positive sentiments of pride and approbation, issue from our concept of self and from our knowledge of good and evil—from the bite of Eve's apple. We blush, and need to, because we evaluate. We rank things as good or better, as bad or worse. And when we evaluate morally, we evaluate ourselves and other selves. Our capacity to evaluate, combined with the knowledge discovered by our sciences and the capacities gained by our technology, places on us the inalienable and portentous burden of moral responsibility.

Accordingly, to affirm that we are natural creatures and then to say no more is to utter a pernicious half-truth. For we are natural creatures with a difference. We have evolved through and past a momentous transformation: the transformation into moral agency. Through our acquisition of articulate syntactic language and our accumulation of culture, we have become self-conscious, deliberative, and thus *responsible* for our behavior. In this sense we are, to the best of our knowledge, unique in this regard. Thus, although we might "retrain" disobedient animals, we do not hold them morally responsible and put them on trial.

The significance of moral agency can scarcely be overstated, for on a planet without moral agents, there are no rights, no duties, no justice, no virtue or vice, and no responsibility, although on a personless planet with a flourishing ecosystem and sentient beings there will be values and value potentials.

With our acquisition of moral agency, we have also acquired the capacity to recognize, celebrate, and enhance nature and thus the responsibility to protect and preserve the natural values around us. We should be ever mindful that with these capacities for recognition, knowledge, and celebration comes the burden of responsibility. For as we come to recognize the value in nature, we also recognize its vulnerability. We are responsible for nature because our science has given us some understanding of the processes at work in nature and our technology has given us the capacity and thus the choice to preserve or destroy our natural estate. And finally, if the central contention of this chapter is correct, through reflection we recognize the values within nature. These four conditions—knowledge, capacity, choice, and value significance—entail our moral responsibility toward nature.

Responsibility, let us remember, is a burden because, given knowledge and capacity, the choice to do nothing is a dereliction. Thus, having taken up the burden of responsibility for nature, we are not morally permitted to set it down again. Returning to the case at hand, we may choose to reintroduce the wolf to the Adirondacks, or we may not. However, knowing what we know about wolves and the Adirondack ecosystem, we cannot opt out of making a responsible decision one way or the other.

To summarize this brief discourse on human nature, we are both natural, along with our fellow creatures, and unique in our possession of the capacities that define our moral agency. The gravest errors in environmental ethics arise from the failure to acknowledge and incorporate both our naturalness and our agency into a system of ethics, to settle for either half of this full truth. On one hand, by "denaturalizing" ourselves we give license to those who would objectify and thus exploit the nature "out there." On the other hand, by depersonalizing ourselves we divest ourselves of moral responsibility for nature, for we thus come to regard ourselves as objects captivated by and helpless in the stream of natural cause and effect.

Morality and Wildness

If my argument has been successful, then both human experience and human nature support my contention that wildness is a condition that enriches personal and communal life. Encounters with wildness can disabuse us of the dangerous conceit that we are nature's favorites and that our time is the culmination of all natural history. Wildness can enhance our personal, moral, and social health by reminding us that we are parts of a larger preexisting order that produced us and now sustains us. From encounters with the wild we can learn and appreciate that all the products of our ingenuity function only as they conform to natural laws and ultimately fail as they attempt to contravene these laws.

On the other hand, despoiling a wild ecosystem diminishes us by reducing our sense of natural place, of perspective, of context. With this we lose our sense of personal transcendence beyond our immediate time, place, and species, turning inward to our species, then to our immediate community, then to our own generation, then to ourselves. As we become narcissistic and alienated, the advantages of the moral perspective and the moral life are lost. We lose this moral vision as we lose our capacity to see ourselves, our species, and our era in their natural contexts—as we forget that we are actors in a drama and participants in an adventure too complex for us ever to comprehend and yet, despite that and even because of that, of ultimate value to us.[4]

To this moral disorder, our scientific knowledge of wild nature, transformed through personal encounter into an appreciation thereof, offers a remedy. From that perspective we can once again regard our world partner with dignity and respect. This perspective, writes Holmes Rolston III, starkly rejects

> the alienation that characterizes modern literature, seeing nature as basically rudderless, antipathetical, in need of monitoring and repair. More typically, modern man, for all his technological prowess, has found himself distanced from nature, increasingly competent and decreasingly confident, at once distinguished and aggrandized, yet afloat on and adrift in an indifferent, if not a hostile universe. His world is at best a huge filling station; at worst a prison, or "nothingness." Not so for ecological man; confronting his world with deference to a community of value in which he shares, he is at home again. (Rolston 1975, 107–108)

Wildness, which once contained our species, now is contained by our civilization and survives only through our sufferance, leading to the false conceit that our artifice can endure and thrive in a totally artificialized world. As wildness becomes ever more rare it becomes more valuable. All further loss of wildness diminishes us, as all recovery of wildness enriches us.

Wolves will reinhabit the Adirondacks only with our permission. Yet that reintroduction may be more valuable to us than to the wolf. For by simply going about their livelihood, indifferent to our needs, the wolves of the Adirondacks would have much to teach us: lessons of our origins, our sustenance, our limitations, and our planetary home. From such lessons as these we just might gain the perspective, appreciation, and motivation to preserve our natural estate, and with it our sustainable place within it.

Thus I conclude, with Thoreau, that "in Wildness is the preservation of the world."

Endnotes

1. By "noneconomic values" I mean values not reflected in market prices, neither actual nor hypothetical (so-called shadow) prices. In other words, noneconomic

values cannot be "commensurated" with cash (i.e., with willingness to buy or sell). Wilderness advocates often argue that enhancing wildness (e.g., by introducing wolves into Yellowstone Park) also enhances economic value by making the area more attractive to tourists. Granted. But this argument plays into the hands of the opponents: What if the cost–benefit balance sheets go the other way? By introducing noneconomic (i.e., nonmonetary) values into the policy equation, I argue that a cost–benefit bottom line against wildness does not suffice to deliver the verdict against wildness. For more about the limits of economic analysis in environmental policymaking, see Mark Sagoff's *The Economy of the Earth* (1988), especially Chapters Two and Three. Also see my "In Search of Sustainable Values" (Partridge 1999); publication is pending, but it is available at my Web site <www.igc.org/gadfly>.

2. Note the qualifiers "*more* wild" and "*less* wild." Because the Adirondack preserve is an ecological island surrounded by human settlements, wolves, like their prey the deer, will also need management. However, a wildlife community in which surplus deer are culled by both natural predation (wolves) and controlled hunting, is more wild than it would be without wolves. To point out just one significant difference, hunters seek out the "best" trophies (e.g., the healthiest bucks), whereas natural predators select the weak. Thus the former degrades the gene pool of the prey population, and the latter, as an agent of natural selection, promotes evolution.

3. This phenomenon is as amazing and mysterious as it is common—as common as riding a bicycle, driving a car, or even walking. As one progresses from the self-conscious struggles of the novice to the automatic, reflexive skills of the intermediate and expert, complex perceptual and kinesthetic awareness fades into the subconscious, thus erasing the perceived subject–object boundary between biker–bike–path, skier–ski–slope, or paddler–paddle and kayak–river.

4. The task of demonstrating moral lessons of wild nature is beyond the scope and the objective of this chapter. However, I have attempted it elsewhere, in an article from which this paragraph is adapted: "Nature as a Moral Resource" (Partridge 1984, 127). In that article I conclude, "We need viable, independent, flourishing natural ecosystems . . . as scientific resources to expand our understanding of what we are biotically and what made us what we are. We need wild ecosystems as economic and technical resources, to provide rare biochemical substances for our future use. We need nature as an aesthetic resource to enrich our sense of delight and wonder. We need natural landscapes and seascapes as psychological resources so that we can put ourselves at ease by returning home again to the environment that made us the natural organisms that we are. And we need nature as a moral resource—as a source of wonder, amazement, admiration, humility, perspective, and solicitude."

Part VI

Thinking Like a Mountain:
Philosophical Analyses of Long-Term
Human Obligations

As we have seen, alongside questions of biology, political process, and cultural beliefs, the issue of wolf reintroduction throws into sharp relief the significance of our present-day decisions for future generations. What should we save for the future? What criteria should we use to decide? These questions are all the more pointed because the proposed restoration seeks a public response to a conscious past decision to extirpate wolves from this and other regions in the United States. One of the aims of Part VI is to take a closer look at the broadened obligations described by authors in Part V. If it is true that public deliberation entails making explicit the values that shape our choices, how should we factor in future-oriented values such as stewardship, sustainability, and biodiversity that have begun to receive broad public recognition?

In Chapter Seventeen, "What Do We Owe the Future? How Should We Decide?" Bryan Norton, professor of philosophy in the School of Public Policy, Georgia Institute of Technology, suggests that the proper question about long-term human obligations is not what future people will need or want, but rather what we, as people of the present, want to save for the future. What are we willing to commit ourselves to in order to project our values into the future? When this is the question asked, the nature of sustainability shifts from a descriptive account of comparative well-being of populations of people across time to a question of how societies constitute themselves as multigenerational moral communities that care for the future and how these communities can project their own values into the future.

In the first section of his chapter, Norton states that our bequest to the future may need to include some specific biological entities. He takes issue with economists who believe that the question of a fair bequest can be resolved at the general, nonspecific level of calculating a fair savings rate. In the second section he seeks a more biologically sensitive criterion for determining what our biological bequest should be. In the third section he observes that placing a high value on biodiversity is not the same as placing a value on a resource. Biodiversity is not a simple sum of existing variety, he says, but a process that generates and sustains multiple evolutionary regimens and thereby creates greater variety across time. In short, biodiversity is a self-generating and self-sustaining force. In the case of the Adirondacks, wolf reintroduction would represent a high-impact, positive effect on biodiversity in North America. In the fourth section Norton clarifies the question of costs and how we should think about determining when the costs of an ecological restoration are bearable. Finally, he discusses how the decision whether to reintroduce wolves into the Adirondacks can make use of these criteria in an open, democratic process, a process in which free people might rationally decide to accept a commitment to reestablish a lost population as an important expression of their legacy to the future.

Chapter Eighteen is by Clark Wolf, associate professor of philosophy at the University of Georgia. In "Sustainability, Environmental Policy, and the Reintroduction of Wolves," he proposes to incorporate biodiversity as a criterion in a definition of sustainability that can guide institutional and individual decision making on questions such as wolf restoration.

Although sustainability is widely acknowledged as an important social goal, says Wolf, there is very little agreement about its meaning. Some hold a "stable endowment" view that the earth is sustainable only if we pass on to the next generation a resource set that is roughly equivalent to the one we inherited. Others hold a "stable productive opportunities" view that we should strive to leave for future generations productive opportunities comparable to those we inherited from previous generations. Productive opportunities are stable when it is possible for future generations to produce the same quantities of consumer goods and other types of goods for the benefit of their members. Still others hold a "stable welfare" view, believing that we should aim to maintain sufficient wealth so that future generations will have the opportunity to achieve the same welfare levels as present generations.

Citing deficiencies in all of these approaches, Wolf offers an alternative view of sustainability as the appropriate basis for social and environmental policy. This view includes two important dimensions. One is a needs conception of sustainability that focuses on long-term human well-being. The second companion conception is based on minimizing human damage to the ecosystems on which human communities depend.

After laying this groundwork, Wolf applies his analysis of sustainability to the question of wolf reintroduction in the Adirondacks. He concludes that the best way to serve the goals of sustainability, natural variety, and the health of the world's ecosystems may be to spend our resources and energies promoting habitat protection and making human needs secure. Thus, a narrow focus on wolf reintroduction should not tempt us to neglect other and possibly more pressing environmental concerns.

Chapter Seventeen

What Do We Owe the Future?
How Should We Decide?

Bryan G. Norton

Our Bequest to the Future

Each generation, either thoughtlessly or thoughtfully, leaves a bequest to generations that follow it. One of the purposes of this book is to ask, Should a restored and flourishing population of wolves in the Adirondack mountains be a part of our generation's bequest to future generations? As my contribution to the analysis of this case study, I intend to focus on how best to understand, formulate, and analyze this question as a question of public values, which, like all important aspects of ecological restoration, must be addressed in an open and democratic public process.

For most of our history as a transplanted civilization, North Americans of European descent have concentrated on two aspects of the bequest they would leave future generations: the moral, spiritual, and cultural aspect and the economic aspect. Early in our history, colonists explicitly stated that it was their purpose to "build a city on a hill" and to set an example of spirituality and righteousness. Today, it seems, most concern for the future centers on building wealth, great cities, and technological prowess—in other words, on maximizing economic growth. Both of these two broad aspects relate to the human heritage—the spiritual and the economic—portion of our bequest. North American nations have, in my view correctly, tried to minimize the role of their government and public actions in shaping the spiritual aspect of intergenerational bequests. To the extent they have succeeded, the religious bequest has become mainly a matter for the private sector. Accordingly, most writers who have examined the question of the nature of our publicly provided bequest—what we should empower our governments to help us accomplish as a society—have focused mainly on the economic and utilitarian aspects of the bequest. That is, they have asked, What represents a fair savings rate? Each generation should add to, or at least protect, the economic capital base they inherited, it is argued. One view, widely accepted among mainstream welfare economists, claims that this requirement exhausts our obligations to the future—that as long as we leave them as well off as we are, economically, we will have fulfilled all of our obligations to the future.

But one of the questions posed in this book—whether our generation might have an obligation, or accept an obligation, to reintroduce wolves into the Adirondack Mountains—may not have an easily calculable economic effect. It has to do with not the economic but rather the biological bequest we leave for future generations. Unlike the methodology of economics, which measures all impacts of decisions in the single currency of units of utility, biology is a science of the particular. In the first section of this chapter, to pursue the idea that our bequest to the future may need to include some specific biological entities, I take issue with economists who believe that the question of a fair bequest can be resolved at the general, nonspecific level of calculating a fair savings rate. In the second section, I seek a more biologically sensitive criterion for determining what our biological bequest should be, and in the third section I consider the exact nature of the resource that is referred to when we place a high value on biodiversity. The fourth section is devoted to clarifying the question of costs and how we should think about determining when the costs of an ecological restoration are bearable. Finally, I discuss how the decision whether to reintroduce wolves into the Adirondacks can be addressed within an open, democratic process, a process in which free people might rationally decide to accept a commitment to reestablish a lost population as an important expression of their legacy to the future.

A Question of Economics?

Let us start by asking how far we can go by casting the problem of wolf reintroduction in economic terms. As noted earlier, most economists and many philosophers believe that the question of what we owe the future is reducible to the question of what constitutes a fair savings rate. That is, each generation must invest enough and wisely enough to expand, or at least not reduce, the total capital of the society. There are some obvious economic impacts of wolf reintroduction. Yellowstone Park and surrounding tourist facilities have experienced a surge of tourists interested in wolves; in addition, there are monetary costs associated with managing the wolf population. The point is not to deny an economic aspect of the decision but rather to examine whether an economic analysis can claim to be reasonably comprehensive with respect to the social values at stake when one generation chooses a bequest for its successors.

The key issue facing economic analysts is that of substitutability between natural and human-built resources (Norton and Toman 1997). In general, mainstream welfare economists have adopted a strong substitutability princi-ple, assuming or arguing that resource shortages are unlikely to affect prices because new technologies allow the development of adequate substitutes for depleted resources. Furthermore, they have taken this high degree of substi-tutability to imply that our decisions about what we owe the future can be guided by principles of fair savings and prudent investment (Solow 1993; Beckerman 1994). They have not been persuaded by arguments that we should differentiate our bequest to ensure that natural as well as human-cre-ated capital exists in sufficient quantities. For them, creating a fair bequest package is not a question of *what* to save (because resources are largely inter-changeable) but *how much* to save (with the caveat that we make good invest-ments and not squander the capital accumulated by our ancestors).

According to the methods and assumptions of welfare economics, biodi-versity should be valued much as we would value any other commodity, but in this case we are purchasing a commodity in the present that will deliver much of its value in the future. To the extent that having wolves in the North-east represents a significant biodiversity value, we must evaluate its impact on future, as well as the present, generations. The economist will tell us that we should protect as much biodiversity as consumers in the present are willing to pay for, and no more. Furthermore, because the basic unit of value in the eco-nomic system is welfare, represented as consumers' willingness to pay (WTP) for a given product, any obligations we have to the future are at most an obli-gation not to reduce their income (their ability to purchase commodities they judge to improve their welfare). In this economic approach to valuing biodi-versity, reintroducing the wolf may prove to have social value if, for example, it leads (as in the Yellowstone reintroduction) to an increase in tourism and economic activity in the area. Or if the wolf were to reduce problems with coyotes, driving them out of some of their recently colonized habitat and reducing their populations, or reestablish some other important ecosystem function, it would be possible to attribute to the newly reestablished wolf pop-ulation a value sometimes called an ecosystem service (Dailey 1997).

Economists also recognize another type of value that they call, somewhat nebulously, existence value. This is the value a person expresses through WTP to maintain or reestablish a species simply to know that it exists. In their research economists have found that respondents express significant WTP to protect such existence values, and this WTP is widely distributed in the Amer-ican public. So one more reason a person might state a WTP to reintroduce wolves, even in competition with other goods they might seek, would be to ensure that future generations will be able to experience wolves and the more whole ecological system that includes wolves. Taking these less obvious values into account as affecting present consumers' WTP for wolf reintroduction, economists in effect reduce concern for the future to a matter of a voluntary, altruistic consumer choice to invest some present income for the benefit of the future (Norton 1991a, 1994; Norton et al. 1998).

This approach to sustainability reduces to what is sometimes called weak sustainability, the doctrine that all we owe the future is the opportunity to be as well off in achieving welfare as we are. In a sense, weak sustainability is the conceptual flipside of strong substitutability. The more one emphasizes strong substitutability between resources as one structures the bequest, the less one will put values on particular items and the more emphasis can be placed on simply making wise investments in purely economic terms. On this thinking, it can be no affront to the future if we use up particular resources, provided we save a sufficient portion of the profits to maintain a nondeclining stock of general capital. According to weak sustainability theory, then, we are obligated only to avoid impoverishing the future economically. Because of fungibility of resources and changing technologies, we cannot identify any particular elements of nature that we are obligated to save. If weak sustainability represents a complete and adequate picture of intergenerational obligations, then it follows that the bequest package one generation owes the next is nonspecific and unstructured. Economic analysis, so used, only measures and compares aggregated wealth acquired and maintained across time, reducing obligations across generations to one of maintaining a fair savings rate.

Why are economists and many philosophical utilitarians convinced by this simplification of the question of what we owe the future? The most common argument in the literature (see, for example, Solow 1993; Passmore 1974) for reducing obligations to the future to determining a fair savings rate rests on ignorance of future preferences (henceforth called the argument from ignorance). It is argued that if we are to determine what, exactly, we should save for the future, we would have to predict or correctly guess what people of the future will want and need. In the words of Nobel laureate economist Robert Solow, "If we try to look far ahead, as presumably we ought to do if we are to obey the injunction to sustainability, we realize the tastes, the preferences, of future generations are something we don't know about" (Solow 1993, 181). Invoking an unquestioned axiom of ethical reasoning—that one cannot be obligated to do the impossible—Solow concludes, because of the ignorance argument, that we cannot be obligated to save particular things that the future might want or need because we cannot predict what they will want or need. This outcome does not bother Solow, however; after noting that economists believe that resources can "take the place of each other," he boldly concludes, "There is no specific object that the goal of sustainability, the obligation of sustainability, requires us to leave [for the future]" (Solow 1993, 181). Maintaining a fair savings rate, he believes, will fulfill any obligations to the future.

In general, economic methods of valuation do not have obvious applications to choices designed to affect ecological function. In his authoritative survey of economic valuation theories and models, Myrick Freeman (1993) agrees with this assessment. At least as currently used, economic methods of valuation have no available methods for evaluating contributory and other ecosystem values. Worse, for reasons explained later in this chapter, there is no apparent extension of standard economic methods to encompass these values. It therefore misses the very important contributory value: the value they confer in keeping the system functioning and contributing the support of other, necessary species. The economic approach apparently places no special value on species or ecosystems, trusting their protection or restoration as a matter for consumer choice. As Freeman says, "If there is no link between [an] organism and human production or consumption activity, there is no basis for establishing an economic value. Those species that lie completely outside of the economic system also are beyond the reach of the economic rubric for establishing value. . . . Rather than introduce some arbitrary or biased method for imputing a value to such organisms, I prefer to be honest about the limitations of the economic approach to determining values. This means that we should acknowledge that certain ecological effects are not commensurable with economic effects measured in dollars. Where trade-offs between commensurable magnitudes are involved, choices must be made through the political system" (1993, 300).

Every species has a value not just as an entity that might be exploited or worshipped; it also has a value that is important to but lost in the functioning of the system at a larger scale (Norton 1987, 1988). Aldo Leopold said it best 60 years ago:

> The emergence of ecology has placed the economic biologist in a peculiar dilemma: with one hand he points out the accumulated findings of his search for utility, or lack of utility, in this or that species: with the other he lifts the veil from a biota so complex, so conditioned by interwoven cooperations and competitions, that no man can say where utility begins or ends. . . . The only sure conclusion is that the biota is useful and biota includes not only plants and animals, but soils and waters as well. (1939, 727)

It is perhaps useful to dwell briefly on this limitation of economic valuation because it illustrates the quandary facing us in evaluating the proposed wolf reintroduction. The limitation in question is that, as noted earlier, economists have assumed a very liberal approach to substitutability between resources and substitutability between natural and human-made capital. This is a substitutability of commodities for each other, however, and the economic conceptualization of environmental value as a collection of commodities suggests no obvious method for counting contributory values of the sort Leopold identified. These contributory values seem to be very important in the long term, affecting future generations significantly, insofar as indirect advantages (such as the improved fitness of deer populations as a result of predation by wolves on the weaker members of the herds) appear and have significance only over multiple generations. This and many other contributory values are important to future generations, but it is difficult to see how they can be measured and accounted for in a model of intergenerational fairness that considers only values reflected in current prices or even in current expressions of WTP.

Solow believes that we have an obligation to be fair to future generations and that we fulfill this obligation by maintaining a fair savings rate, avoiding the impoverishment of the future. How far can this model of intergenerational obligations carry us in specifying what we should save for the future? To maintain a fair savings rate efficiently, Solow could argue, we are implicitly obligated to make good investments with the portion of our income that is set aside as a contribution to fair savings for the future. And if any given element or process of biodiversity has high contributory value, it seems to follow that we should invest in it. Solow thus accepts, by implication, an obligation of present consumers to invest wisely for the future, and if a wolf reintroduction is a wise investment, people in the present should make this investment as a contribution to savings—to wealth creation. If economic actors in the present choose their investments wisely, Solow could say, then today's prices should reflect in present dollars the future value of specific objects such as elements and processes of biodiversity. As an economist who relies on market analysis, Solow seems on strong ground here because it is an assumption of market-based economic analysis that consumers act with full knowledge of the consequences of their actions, so we should be able to expect today's consumers to make the investments that will maintain capital efficiently and avoid impoverishing the future.

So far, this position seems compatible with a protectionist attitude toward biodiversity and nature: Present markets reflect future values, so our obligation is to protect for the future the elements and processes that hold open the most and the best options for future consumers. Solow faces a serious quandary, however, that is exacerbated by his appeal to ignorance as a reason to limit our concerns to maintaining a fair savings rate in the first place (Norgaard 1990). He cannot on one hand claim that prices correctly reflect the future value of increments in biodiversity protection (on the grounds that today's informed consumers' choices reflect future values) and, on the other hand, claim that we are too ignorant of future needs and wants (demands) to define specific obligations to save specific elements and processes of biodiversity. In doing so, he would be implicitly imputing to today's consumers knowledge that he has claimed nobody can have. Either present prices reflect future demands and, therefore, prices provide a good guide to investments in our bequest for the future, or they do not. If they do, then Solow is necessarily wrong in saying that we do not, and cannot, know what the future will want or need because it is precisely this knowledge consumers will need to predict what will be good investments if they are to maintain a fair savings rate for the benefit of the future. If today's prices do not reflect future values, however, then today's prices cannot be considered a good guide in the present to investments for the long-term future.[1]

Either way, we must conclude that there is information that is essential to know, information that is not reflected in analyses of present prices and fair savings rates. Weak sustainability cannot, without supplementation, without biological information and information about future values, guide us toward good investments in the well-being of the future in policy areas such as biodiversity protection and restoration.

To summarize the argument so far, on the economists' assumption of ignorance of future demands and full substitutability between resources, there is one and only one way in which the present can harm the future, and that is by impoverishing future consumers by spending more than we produce. But this seems too narrow a concern in cases such as restoring particular species because it seems reasonable to posit a closer relationship between maintaining a diverse habitat as our environment and our maintaining real options that will be valued and valuable in the future. If we want to protect the future from loss of indirect but highly significant values such as the contributory values of species and processes, we must supplement economic valuation with noneconomic principles and reasoning. Despite the apparent inadequacy of weak sustainability to provide a comprehensive accounting of the requirements for fairness to future generations, it nevertheless seems reasonable to consider weak, or economic, sustainability to be a necessary condition for sustainable living. But the argument of this section shows that if we want to articulate a more structured and specific set of obligations to guide the formation of a more structured and biologically sensitive bequest package, we must consider weak sustainability as only a necessary but not sufficient conceptualization of the intergenerational trust.

An Ecologically Informed Safe Minimum Standard Criterion

Some environmental economists have recognized that weak sustainability based on rational, individual consumer choice does not go far enough in providing a presumption in favor of saving specific productive resources such as biodiversity. A number of respected economists and others have advocated an important modification of cost–benefit accounting (CBA) by applying the safe minimum standard (SMS) rule of conservation. This rule says that a productive resource should be saved if the costs are bearable (Ciriacy-Wantrup 1968; Bishop 1978; Norton 1991b; Farmer and Randall 1997). In effect, this modification assumes that the resource is worth saving (thereby avoiding the need for a benefits analysis) and focuses analytic attention on the cost side of the ledger, encouraging a search for affordable means to an unquestioned goal. In effect, this modification shifts the burden of proof to those who would eliminate a productive resource. To sacrifice a species, for example, SMS advocates believe one must make a case that protecting the species will impose unbearable costs on society. This modification is an important step in the right direction, as I have argued elsewhere (Norton 1987, 1991b).

One advantage of the principle is that although the particular term *SMS* has emerged from the discipline of resource economics, it has an important rough analogue in political jargon, especially in Europe and Australia, in the idea of a precautionary principle. The precautionary principle states that in situations of high risk and high uncertainty, always choose the lowest-risk option. I prefer the SMS criterion to the precautionary principle because the SMS provides explicit direction in determining when risks should be avoided. SMS instructs us that in situations of high uncertainty and risk of irreversible impacts such as species extinctions, adequate steps should be taken to protect the vulnerable resource if the costs of protection are bearable.

It cannot be denied, however, that the SMS criterion suffers from vagueness in both of its crucial terms: *resource* and *bearable costs.* How can we decide what we should save or restore? Of course, one can say, "Save everything! Restore everything!"[2] But that policy is not possible without unbearable costs, not to mention the fact that, taken literally, this injunction would freeze nature unnaturally. So, if we are to apply the promising SMS criterion, we must specify which particular resources should be given priority in protection and recovery programs.

The rule that we should save a resource is especially vague in the case of biodiversity because one could emphasize individuals, populations, species, and ecosystems at several scales.

Can we give biological or ethical reasons to determine what to save and what to restore? Can we give reasons that are clear and convincing enough to persuade the people of New York that they should take steps to reestablish populations of wolves in feasible wilderness areas? Can biology provide some general guidelines as to what should be in our bequest? The question of what to save as a biological bequest is addressed in the next section; the question of how to clarify "unbearable costs" is the subject of the final section. Taken together, the arguments of these two sections should allow us to apply the SMS criterion with greater precision.

Biodiversity and Resources

If the reader will permit the interjection of a personal note, I have been puzzling over, reading a lot about, attending conferences on, and writing a good bit about the value of biodiversity since 1980. I thought I had heard it all. About 2 years ago I read an article, before its publication, that changed the way I think about this much-discussed topic. The author of the article, Paul Wood, first departs from usual practice and distinguishes sharply between biodiversity and biological resources. According to Wood, biodiversity simply equals "the differences among biological entities" (Wood 1997, 254). He then makes a persuasive case that it leads only to confusion to consider biodiversity to be a resource among other resources and to trade off elements of biodiversity against other resources. Biodiversity, he argues, is not a resource among others, but a generator—a source—of biological resources. As such, biodiversity is a necessary condition of enjoying biological resources, especially over extended time. To use an analogy from the field of economics, biodiversity should be valued on the higher logical plane on which economists and some policy analysts value free markets as generators of value. Just as economists would find it odd to be asked, "What is the market value of a free market?" ecologists may find it odd to think of biodiversity as a commodity. Free markets, on the theory of the famous invisible hand by which individuals seeking their own interest create an efficient system for all, generate value. Similarly, advocates of biodiversity do not conceive biodiversity as one resource among others, to be traded off to get an "optimal mix" of biological and other resources, but rather as a source that, in the long run, is a necessary precondition for the system to generate more biological resources. This generative process exists on at least two levels. On one level, diverse ecological processes provide diverse outputs, including standing resources of many types and a variety of ecological services. But on a longer scale, diversity of biological processes, especially at the ecosystem scale, is also responsible for generating and maintaining diversity itself.

To see the importance of diverse biological processes, it is useful to examine, briefly, an often-used working definition of biodiversity. Biodiversity is often defined as the sum total of all diversity, including the genetic diversity within species, the diversity of species, and the diversity of habitats within which species live. Though perhaps useful as a simple definition for the purposes of "accounting," this definition, by listing the diversity of entities at important biological levels, does not sufficiently emphasize the importance of diverse processes because it concentrates on the collection of diversity at a single point in time. If one thinks of biodiversity not as a simple sum of existing variety but as a process that generates and sustains multiple evolutionary regimens, and hence creates greater variety across time, biodiversity is a self-generating and self-sustaining force, as close as nature has come to the much-sought *perpetuum mobile*. Systems that are diverse provide more opportunities for species to create and evolve into new niches. Diversity creates more diversity (Whittaker 1969; Levin 1981; Norton 1987; Wood 1997). Conversely, losses in diversity eventually lead to more losses, as species that depend on lost species are themselves extirpated, and a cascade of losses is entrained. Adding a time element to our thinking about biodiversity, in this sense, shifts us away from thinking about biodiversity as a list of entities or resources and toward an emphasis on the creative, self-sustaining nature of diverse ecological systems.

To continue with this level-oriented or scale-sensitive treatment of biodiversity, we can now examine the relationship between multiple evolutionary processes at different scales and the traditional understanding of types of biodiversity. Writing in the 1960s and 1970s, R. H. Whittaker (1969) usefully separated three types of diversity in ecological systems. First, there is the diversity of species, sometimes called richness, in any given habitat. On this level, for example, we know that tropical rainforest habitats are much more diverse than are tundra or even temperate habitats. But the diversity within habitats probably is less important to generating and maintaining diversity than is the diversity of habitats that exist in an area. This cross-habitat type of variety is important in providing multiple opportunities for species to adapt and create new niches by providing in close proximity a variety of alternative habitats, which encourages opportunities and thus more variance. If a species can develop a new survival trait, it may be able to colonize a new type of habitat and, once it has colonized a new habitat, the species is submitted to a whole new set of evolutionary pressures and may adapt in quite new directions. In the exceptional case, which usually occurs when the new population is largely isolated from its earlier habitat, a new species may emerge. The total diversity of an area at any given time is a product of within-habitat diversity and cross-habitat diversity. But total diversity over time is a function of the diverse processes at all scales because these diverse processes provide and maintain opportunities for diverse species and populations to survive and adapt across time. Habitat diversity provides diverse options for survival, and each of these diverse options may represent an opportunity for some organism with a genetic or behavioral quirk. Similarly, this occurs for populations.

It is now possible to explain, in broad biological terms, why a wolf reintroduction in the Adirondacks would represent a high-impact effect on the biodiversity of North America. Because most areas of the continental United States no longer have wolves as their top predator, the Adirondack system would be unusual in this respect; because the top predator exerts evolutionary pressure that is felt throughout the community, a variety of old and new relationships will be reestablished or established. At least in Yellowstone, the return of wolves apparently has caused a greater diversity of wildlife (Robbins 1997), and a similar impact may occur in the Adirondacks. However that comes out, the Adirondacks take on added biological interest in their own right in that they increase their difference from other systems of the northeastern United States and thus contribute to the total diversity of the region.

I believe that the goal of biodiversity management at the continental level should be to maintain and restore the greatest authentic total diversity possible (Norton 1987). Calling the goal "authentic" total diversity adds to the idea of maximizing total diversity the additional requirement that species counted must either have existed historically or could reasonably be expected to colonize without the aid of human managers. This requirement rules out augmenting diversity by importing exotics, emphasizing that the maintenance of diversity should favor traditional species because these have evolved to have competitors and symbionts in the habitat.

To be biologically precise, the resource to be protected should be the source of biological resources lodged in the differences between biological entities at all levels of the biological and evolutionary hierarchy. The policy goal should be to protect biodiversity, understood as the sum of all biological differences at all levels and scales, including especially the diverse processes that underlie, maintain, and augment the diversity spiral. These differences between entities, including differences in the evolutionary trajectories they embody over time, sustain diverse processes and diverse entities that, as Wood argues, are the source of biological resources.

Reintroducing wolves, the authentic top predator for the region, to a special area such as the Adirondacks would reestablish processes that support differences between entities in the Adirondack system. Looking at the larger scale of the eastern United States, total diversity likewise will be augmented and sustained. The existence of a more complex and authentic area within the larger region is sure to expand the number of evolutionary trajectories for many species, creating more diversity at every scale of the system. This increase in diverse evolutionary trajectories leads to greater and greater cross-habitat diversity, including especially an augmentation of cross-habitat diversity as a potent driver of the processes creating more total diversity. Diversity creates more diversity. Losses of diversity lead to future losses. This self-sustaining and self-creating process is ultimately the source of all biological resources.

I began this section by asking what exactly is the resource in question in the SMS criterion "save the resource, provided the costs are bearable," as applied to biodiversity. The answer, I have argued, is that the resource in question is the sum total of differences between biological entities at all levels and that the sum includes, very importantly, differences between ecosystem processes because these ultimately are the source that must be saved. These processes create and sustain both the differences themselves and the flow of biological resources that are necessary to fulfill human needs. But even this may command us to save too much; it is no doubt impossible to save or restore all biological differences. Why wolves? Why in the Adirondacks?

My answer is that reintroducing an authentic top predator to a unique habitat such as the Adirondack Mountains will greatly multiply cross-habitat differences in the larger region. This particular reintroduction would recreate an old evolutionary dynamic, diversify the evolutionary regimens faced by many species in the eastern United States, and create a habitat that exists almost nowhere else in the region. If one wants to maximize our bequest of biodiversity, understood as the biological differences that are the source of biological resources, the best investment available would be to restore the authentic top predator to an area within a larger region where it once existed, regaining lost cross-habitat diversity and spinning the flywheel that creates and maintains diversity over time.

Interpreting "Unbearable Costs"

Though helpful in determining the target of the effort of SMS advocates to save the resource, these considerations do not fully resolve the ambiguity of the other key phrase of the SMS criterion, "unbearable cost to society." The argument that wolves can recreate lost cross-habitat diversity shows that a great social value is at issue, but it remains a difficult question: How might the supporters of a reintroduction in the Adirondacks make a case to the residents of the park that diversity is valuable to them and to all humanity and that the costs of such a reintroduction are bearable, given the benefits to be expected?

The first and in some ways most important answer to these questions is that they will demand political answers. In a democracy, the political will necessary to undertake a restoration must emerge as a legitimate expression of the community's will. In the case in point, the decision to reintroduce wolves must have strong support at every political scale, from local to statewide, especially including residents of the park, if it is to succeed. We have recognized a significant gain to the future—the augmentation of cross-scale diversity—but can also recognize that protecting this increased value (which will emerge over generations) might be costly to at least some members of the present generation. But now the advantages of using a scale-sensitive model for expressing values may help. It may be possible that the biological bequest can in this case be protected with little cost to the future if we can identify independent dynamics that generate these values (Norton and Ulanowicz 1992). It may be possible, with creative planning and policy, to create a win–win situation by finding a path toward community development that will generate short-term economic growth in a way that is compatible with protecting and augmenting biodiversity.

I do not know whether the political will to reintroduce wolves exists at these various levels, nor do I know whether it could be created or developed. This is yet another fascinating question, but one that can be answered only by observing the political process unfold over the next years and decades. Rather than speculate about what will happen, a question that can only be answered empirically as the future unfolds, I will ask another question: On what scientific and moral basis might a human community, through a process of deliberation and participation, decide to accept a moral obligation to protect and augment authentic biodiversity? In answering this question, I will use my remaining space to explain why diversity can be expected to provide social value, even if uncertainty and ignorance of future preferences make it difficult to identify particular uses or commodities of units of biodiversity. If the residents of the park agree that wolves, as top predators, are an important aspect of the diversity of wild places, then my addition of an explanation of how diversity can be expected to have social value should strengthen the argument that a reintroduction, if biologically feasible, should be a social priority.

Again, it is helpful to keep in mind the multiscalar nature of biodiversity and the values it supports. By arguing that reintroducing wolves to the Adirondacks increases the cross-habitat and thus the total diversity of the eastern United States, it becomes clear that one important social value affected by the proposed reintroduction—protecting total diversity for future generations—is manifest at a larger, regional and national scale. This recognition also helps us understand the political complexities of the wolf reintroduction process. Local people, who look mainly at values to their community, might conclude that from their point of view, the reintroduction is not worth the costs and risks involved. But when the multiscalar nature of biodiversity is understood, it is also possible to see that from the larger scale of the region or nation, existence of wolves in this unique habitat may have enough value to outweigh the loss and risks to the local community.

This conceptualization of the problem suggests that, to the extent that augmenting diversity is a social or public good, it ought not to be imposed on individual property owners, and the society may have an obligation, based on intragenerational fairness, to compensate property owners and the communities that may suffer losses as a result of the restoration effort. The Defenders of Wildlife compensation fund is one such method. If the greatest values of biodiversity are experienced over generations and at a national scale, the costs ought not to be inordinately heavy at the local scale. It may be necessary to provide compensation in some form to local communities or some groups that may suffer inordinate losses. Let us then assume that for the most serious losses suffered at a local level, a compensation plan that is fair and acceptable to local communities could be worked out in a process of negotiation.

Having acknowledged that some compensation may be due landowners and residents if they are supporting a public value, this places the local communities at the center of the decision process, a process that is embedded in a comprehensive, multiscale examination and evaluation of the reintroduction at a regional and state level. Assuming intragenerational equity concerns can be addressed and that the feasibility studies are positive, we can now concentrate on characterizing the long-term benefits of reintroducing wolves to the Adirondacks and the costs to the future if we fail to make this investment when we still have a chance for it to succeed. We can at last ask whether the biological importance to future generations of augmented diversity, understood as total diversity, justifies our generation in including restored wolves in the Adirondacks as an important part of our bequest to future generations.

But immediately upon asking this central question, we encounter our old nemesis, ignorance about the preference of future people. The question of specifying bearable costs turns our attention to an analysis of benefits, but how can we understand benefits to future people if we don't know what they will want or need? One thing we can say, based simply on our analysis that favored the SMS over CBA as an appropriate decision tool, is that we are not as ignorant of the preferences of the future as Solow suggests. It seems obvious that people of the future will not want to discover toxic time bombs in our waste repositories, and they will prefer a diverse and opulent world of living things to a sterile, dead planet. In fact, we have a pretty good idea what will be valuable—and what disastrous—for future people and future communities.

Once again, our understanding of biodiversity as the differences between biological entities and processes provides guidance. We value differences and the varied biological entities and processes that embody and support them because these differences support and enhance human options in the future. Thus, although we probably lack specific guidance about wants and needs, we do know that people of the future—assuming they will value free choice as we do—will want to have options and opportunities available. In addition to impoverishing the future, we could be unfair to them if we unjustifiably reduced their options, their range of choices. We should protect differences between biological entities when, and to the extent to which, they are necessary to hold open options that may prove important to people in the future. Although much more must be said to clarify this point, I am suggesting that the SMS can be supplemented and made more biologically specific by relating biological differences to the social value of maintaining freedom, which is accomplished by protecting resources associated with important options and opportunities (Norton 1999). The SMS might stand in addition to the requirements of weak sustainability as a second necessary condition of sustainable living, especially applicable to decisions that strongly affect the future of irreversible ways.

We have articulated, in broad outlines, a two-criterion basis on which we can evaluate possible options and policies with respect to their sustainability and fairness to future generations. To be fair to the future, our actions today must avoid impoverishing the future (as measured by economists' weak sustainability notion of simply maintaining a fair savings rate) and avoid destroying options that may prove important in the future. I am hypothesizing, then, that a general and conceptual connection holds between protecting biological differences across generations and maintaining freedom of choice.

It is tempting to say that saving options for the future is valuable no matter what the future turns out to want or need, and in one sense this is true (to the extent that options support freedom of choice, they are generally valued); however, we may still not know which options are worth investing in, given that we probably will not be able to protect or reestablish all authentic biological differences. This limitation is not debilitating to our use of the biologically informed SMS in the case of the wolf reintroduction, however, because we have a strong biological reason to favor this particular reintroduction. This reintroduction would recreate a process difference that supports and reestablishes many other differences in competitive regimes between species and hence would have a large net positive impact on differences and, accordingly, on options and choices of future humans.

However, it may be more difficult to provide a general account of which differences to maximize if the goal is to maintain important, specific options for the future. Here, our theory may be silent. Beyond emphasizing the importance of protecting sources of biological difference—differences that clearly have high contributory value—it cannot provide specific guidance about all the options that should or must be saved for future generations if we are to be fair to them. Beyond a general value placed on contributory values, it would still be helpful to know which options we should protect for the future.

Once the obligations to protect major sources of differences is fulfilled, it will be a matter of deciding, in each specific geographic place, which options and which associated biological differences should receive priority efforts. To decide specifically which biological differences to protect in a particular place, we seem once again to be stymied by the problem of ignorance. What will future residents of this place want? What should we save for them? As a way through the ignorance-of-future-preferences impasse, I propose a bold solution, a solution that shifts the local decision context radically. Let us not ask, What will the people in the future prefer? Let us instead ask, What values and opportunities are community members in this place willing to commit to and work to protect? In addition to the weak sustainability obligation to not impoverish the future, I am suggesting that the second requirement of intergenerational fairness is to identify and protect the differences and processes that represent the values of the community in question.

In effect, this solution is to identify a sociocultural process by which communities might identify important options, options associated with their very sense of the place they call home. We can call these values constitutive values, and the options and opportunities associated with them are constitutive options. Constitutive values for people are values such that, if they were lost, the person in question would no longer call that place home and would judge that its integrity had been compromised. Suppose we could describe a public process that truly engages the public, at a given geographic place, in an iterative and ongoing search for a sense of place and associated values and that a consensus emerges about certain values as constitutive values. Suppose, further, that this process included all involved stakeholders, allowed open discussion, and resulted in a set of values that are widely thought to be constitutive of the community and its character. Accordingly, there could be public discussion of the particular options and the associated biological differences that give distinctiveness and meaning to the place. If these steps were all completed, then I think one might consider these constitutive values, accepted through democratic processes, as the democratically expressed positive values that tie the community to the specific place.

All of these exercises, I submit, can be thought of as exercises in community self-definition. This approach owes much to the conservative philosophy of Edmund Burke (1790, 93–94), who stated that a society is "a partnership not only between those who are living, but between those who are living, those who are dead, and those to be born." What we add to Burke's historical conservatism with respect to institutions is an ecological sense of the ongoing, multigenerational community as anchored in a physical place. If a community accepts responsibility for a place and for the values that serve as threads weaving the multigenerational tapestry of life and community into a single distinctive culture and community in that place, then it would be reasonable for the people of that community to protect and preserve the constitutive links between their culture and their physical place.

To operationalize this advice, a community, mindful of its natural and cultural history, might pose the question as one of choosing a development path that is fair to the future. Here, we understand a development path as a way in which a community would develop from a given point in time. So far, we have proposed two general criteria by which we can judge the fairness of the proposed development paths. We can then evaluate development paths according to these general criteria: a criterion of economic growth and fair savings (measured as short-term economic impacts on the total capital base of the society, or weak sustainability) and a second indicator or suite of indicators that are designed, after public discussions of values and associated options, to track democratically accepted and supported constitutive values. Because the weak sustainability criterion tracks individual decisions conditioned by today's market, this criterion reflects today's prices as they are set in today's markets. The second criterion, which demands that we protect and enhance authentic biological and ecological differences, applies at a multigenerational scale and tracks the ecological and cultural character of the place over the longer run. I have argued that the added diversity and unique wildness reestablished by a wolf reintroduction would maintain and enhance important opportunities for future generations. These long-term values should be treated on the model of a trust by which each generation invests in a series of community-defining decisions, decisions that will determine the identity and character of the place the community inhabits. Again, the emphasis on protecting options applies at multiple levels, and some system-level decisions and impacts (such as extirpations and reintroductions of top predators) may be very important to the future.

Communities in the Adirondacks, the State of New York, and the larger, communities of New England, on this line of reasoning, are obligated to reintroduce wolves if and only if this policy emerges from a process of community self-definition. I see the role of the Adirondack Citizen's Advisory Committee as exemplifying and perhaps directing this process.

Much has been written about the "science" of sustainability—and surely science has a role in the search for sustainability—but my line of reasoning suggests that the most important aspect of identifying a sustainable development path is an act of community commitment. One might say, to borrow a useful term from philosopher J. L. Austin, that what is involved is a performative act, an act of commitment by a community that is willing to embrace and project certain values into the future. Austin notes that many apparent acts of description actually function as performatives. As examples, Austin mentions "I do" (when uttered in the context of a marriage ceremony), "I name this ship the Queen Elizabeth" (while striking the ship's bow with a bottle of champagne), and "I bequeath this watch to my brother" (in the context of a will). He then says that in these examples, to utter the sentence in question (in the appropriate circumstances) is not to describe the doing of what is being done, but rather to do it (Austin 1962, 5–7). He proposes that we characterize such uses of language as performatives, and mentions that they can be of many types, including contractual and declaratory. Later, Austin says, "A great many of the acts which fall within the province of Ethics . . . have the general character, in whole or in part, of conventional or ritual acts" (Austin 1962, 19–20).

Applying Austin's idea, and based on the analysis of this paper, a new way of thinking about intergenerational morality and sustainable living emerges. If we see the problem as one of a community making choices and articulating moral principles—a question of which moral values the community is willing to commit itself to—then the problems of ignorance about the future become less obtrusive. The question at issue is a question about the present; it is a question of whether the community will take responsibility for the long-term impacts of its actions and whether the community has the collective moral will to create a community that represents a distinct expression of the nature–culture dialectic as it emerges in a place. Will they rationally choose and implement a bequest—a trust or legacy—that they will pass on to future generations? This approach bypasses the problem of ignorance because we need not ask what the future will want or need. Rather, we ask by what process, and on what grounds, a community might specify its legacy for the future. If one wants to study such questions empirically, important information is available. For example, one might study how communities engaged in watershed management processes or community-based watershed management plans achieve (or fail to achieve) consensus on environmental goals and policies. Although empirical studies such as these may contribute to the process of community-based environmental management, I am suggesting that the foundations of a stronger sustainability commitment lie more in the community's articulated moral commitments to the past and to the future than to any description of welfare expectations or outcomes.

This basic point makes all the difference in the way information is used to define sustainability, and it changes the way we should think about environmental values and valuation. So we face the prior task—and I admit it is a difficult and complex one—of developing community processes by which democratically governed communities can, through the voices of their members, explore their common values and differences and choose which places and key values will be saved, achieving as much consensus as possible, and continuing debate about differences. These commitments, made by earlier generations, represent the voluntary, morally motivated contribution of the earlier generation to the ongoing community. The specification of a legacy, or bequest, for the future must ultimately be a political problem, to be determined in political arenas. To choose to protect wolves, on this view, would be to articulate and, through a group process, affirm certain values as constitutive of the distinct and unusual wildness of the Adirondacks.

We must also realize that this choice need not conflict with good policies for economic development. If my argument that specific dynamics are unfolding at different scales is correct, it may be possible to turn controversy into consensus by showing that a reintroduction policy will have positive economic impacts as tourists visit the area and create economic opportunities for entrepreneurs, providing short-term benefits that increase the standard of living of citizens of the Adirondacks and at the same time set in motion the reestablishment of processes that will augment and sustain diversity. The policy would then yield benefits on economic, utilitarian criteria (manifest in the short run) while representing an investment in the options and opportunities that will be available to future people. I conclude that if all of these conditions were fulfilled and if a multilevel political process were to lead to a consensus to protect the future's diverse biological heritage and the opportunities it supports, then our bequest to the future should include a restored population of eastern timber wolves in the Adirondacks.

Endnotes

1. This argument is an elaboration of reasoning first developed by Norgaard (1990).
2. See Howarth (1995) for an argument that the future has proprietary, moral rights in resources, which may imply a policy close to that of total preservation.

Chapter Eighteen

Sustainability, Environmental Policy, and the Reintroduction of Wolves

Clark Wolf

During the early history of the United States there must have seemed little reason to provide safeguards for preserving forests and ecological systems. At that time, human encroachment seemed insignificant in comparison to the size of the continent. But by the late nineteenth century it was clear that the virgin forests were finite. It was abundantly clear that development in general and logging in particular had changed the nation irrevocably. Recognition that natural resources and wildlands themselves are finite was a spur that led to the creation of many of America's parks and preserves that date from this period. In 1894, a prescient New York State legislature created a 6-million-acre park in northern New York, stipulating in the state constitution that 2.5 million acres of that park would be "forever kept as wild forest lands." The creation of the Adirondack Park was a remarkable effort to limit the effects of economic growth. National and state parks such as the Adirondack Park serve a variety of social purposes, including commercial exploitation as well as recreation and environmental preservation. But they are widely viewed as preserves where the forces of economic development are held at bay and where, at least locally, we can strive to maintain the integrity of natural areas for the future. In a social context marked by change, instability, and unsustainable practices, there is hope that at least in these islands of wildlands we may be able to preserve small but sustainable pieces of the forests, plains, and deserts that recently covered the continent.

Many people share a desire that these islands of wilderness should be left alone or restored to their original condition as much as possible. In this interest, human structures often are dismantled or left to disintegrate. And it is in the service of this goal that many environmentalists have worked to prevent the extinction of native animals and plants and to reintroduce those that have been exterminated as a result of human activities. Such reintroduction efforts are under way in many places and have sparked controversy but also overwhelming support. The nation witnessed the reintroduction of the California condor and the release of ospreys in the Northeast. But no reintroduction effort has been more controversial or more closely watched than the reintroduction of wolves in Yellowstone Park. Schoolchildren have followed the progress of reintroduced wolves, and many people have gone to the park in the hope of seeing wolves or hearing them at night.

Reintroductions present many practical, concrete problems. But in evaluating reintroduction policies it is valuable to consider them from a broad perspective so we can understand them in the context of other goals. In this chapter I offer criteria for evaluating reintroduction efforts, specifically the reintroduction of wolves in the Adirondacks. I do this by situating such reintroduction policies within our broader environmental aims.

For many environmentalists, concern for protection, preservation, and restoration are inextricably connected with concern for future generations of human beings who will inherit the world we leave behind. This does not mean that such environmentalists want to protect wildlands merely for their instrumental value; indeed, if wildlands were not valued for their own sakes, it might be difficult to explain why we should hope that future generations might take an interest in them. One of the things we might want to pass on to future generations is the capacity to appreciate wildlands and integral ecosystems for their own sake, not merely as resources to be exploited. We might do this by sharing our own appreciation with our children or working to influence policy and institutionalize environmental protections in law. Beyond these remarks, this chapter does not discuss whether ecosystems and wildlands have intrinsic value, but it should be clear that I am not implying that their value is merely instrumental.

I begin with a discussion of sustainability, which I take to be an important goal, perhaps the most important goal for environmental policy. Whatever our fundamental reasons for environmental protection and restoration, our pursuit of these aims will be effective in the long run only if human institutions can become, in the appropriate sense, sustainable. After discussing different senses of sustainability, I consider the relationship between sustainability and biological diversity and the role that might be played by reintroduction efforts as a means to promote sustainability in our relationship with the earth's ecosystems. Finally, I consider a diverse set of considerations relevant in evaluating reintroduction programs in general and the proposed reintroduction of wolves in the Adirondack wilderness in particular.

Sustainability, Intergenerational Justice, and Biodiversity

There is wide agreement, at least among environmentalists, that sustainability is a worthy ideal if only we could achieve it. But there is little agreement about what sustainability is. The ambiguity of the term has concrete implications, for different conceptions of sustainability support different policies, and surface agreement about sustainability as an ideal may mask fundamental political and ideological disagreements. In the next section I consider a series of different sustainability concepts and argue that we should take at least two of these concepts as normative, that is, as appropriate objectives or aims for social and environmental policy. These are a needs conception of sustainability, which focuses on the long-term well-being of human beings, and a companion conception based on minimizing human damage to the ecosystems on which human communities depend.

We leave our descendants a mixed legacy. Parts of this legacy are positive: knowledge and information, works of art, roadways and communications systems, and a network of legal, social, and cultural institutions. But it is obvious that parts of our legacy to future generations are negative and perhaps even tragic: We leave behind social and ethnic conflicts that could flare up to war even after centuries of dormancy. We leave depleted oil and mineral reserves. Our efforts to extract these resources have left huge scars on the land, and we continue to inflict enormous damage to the ecosystems on which we depend. Recent studies indicate that California's wilderness is losing nearly 100 acres each day to logging, mining, and other human intrusions, and that state forests there have lost 625,000 acres in the past 20 years (California Wilderness Coalition 1998). Current estimates suggest that fully 10% of the world's coral reefs have already been destroyed, mostly in the last 20 years. Projecting present rates of destruction into the future leads to the prediction that 70% may be gone within the next 20 to 40 years (California Wilderness Coalition 1998). But there is every reason to believe that the current rate of destruction will not be maintained; it will *increase* as human exploitation of marine ecosystems becomes more efficient.

Most disturbing is the general rate of extinction of species worldwide. The earth is experiencing a sixth great spasm of extinction, the most dramatic since the Cretaceous era 65 million years ago. Although there is still discussion among scientists about the causes of the first five of these events, there is wide agreement that humankind has caused the current extinction rate to jump. Niles Eldredge (see Chapter Twenty, this volume) reports that the most reliable scientific estimate suggests that we are losing 30,000 species each year—more than three every hour. The losses have been greatest for large mammals, amphibians, and beetles, but the effects been felt in all parts of the globe and by almost all species. It is certain that the lives of our descendants will be changed by this massive extinction. It seems clear that the damage and destruction of so many of the world's biological systems will be worse for them, even if we imagine that we compensate for these losses by developing new technologies or increasing the size of the capital stock on which they may draw. It is not clear just how their lives and opportunities will be changed by current damage to ecosystems or current biodiversity loss. It may be that the human benefits of environmental destruction will last for another hundred years as we exhaust the terrestrial resources of our planet and become more adept at mining its marine resources. But when human activities cause damage at a rate faster than the earth's ecosystems can mend themselves, our activities are unsustainable in an important sense.

Alternative Conceptions of Sustainability

Stable Endowments

What does it mean to say that human activities are unsustainable? Perhaps the most common conception of sustainability is the stable endowment conception. On this conception, our treatment of the earth is sustainable only if we pass on to the next generation a resource set that is roughly equivalent to the one we inherited. But stated in this way, the stable endowment conception is ambiguous and perhaps impossible to meet. Mineral and petroleum resources are nonrenewable: If we use any at all, there will be less for future generations. And it is often argued that the benefits we pass on to future generations may compensate them for the reduced stocks of nonrenewable resources. More importantly, the stock of resources we pass on to future generations may be less important to them than the opportunities those resources represent. Technological and social change may imply that future generations will value different resources than the ones on which we rely. For example, it is sometimes argued that petroleum resources should be used to facilitate a transition to less destructive, renewable sources of energy. Perhaps our continued use and depletion of these nonrenewable resources is necessary for such a transition—not that current depletion is actually serving this aim—but it seems to be forbidden by the stable endowment conception. If the transitional use of nonrenewable resources were necessary for long-term environmental protection and for the long-term satisfaction of human needs, it would seem strange to adopt a conception of sustainability that would prohibit us from taking the necessary steps to move toward reliance on energy sources that would serve human needs without inflicting such severe environmental damage.

Stable Productive Opportunities

Considerations such as these have led some to articulate a different sustainability standard, which might be called the stable productive opportunities conception. On this conception, advocated by political scientist Brian Barry (1989), we should strive to leave for future generations productive opportunities comparable to those we ourselves inherited from previous generations.

Productive opportunities are stable when it is possible for future generations to produce the same quantities of consumer goods and other types of goods for the benefit of their members. So, on this conception of sustainability, depletion of nonrenewable resources is sustainable as long as we offset the cost to future generations with technological advance so that future generations will not have fewer opportunities than we have enjoyed. But this conception has some obvious problems. First, some goods may be nontradable; there may be nothing we can provide for future generations to compensate them for the destruction of the Mississippi delta or the benefits of a breathable atmosphere. Similarly, there may be nothing we can do to compensate for the loss of biodiversity caused by present human activities. More importantly, the notion that we should leave future generations undiminished opportunities fails to accommodate the fact that future generations may be much more populous than we. As human population increases, present production rates will soon be inadequate to meet its needs. Indeed, there is good reason to believe that they are already inadequate. Where population is growing, stable productive opportunities may leave all members of later generations worse off than their predecessors. Surely it will seem strange to regard such steady decline as sustainable in any normative sense.

Stable Welfare

Such considerations might make a stable welfare conception of sustainability attractive. Such a criterion would recognize institutions as sustainable when they provide future generations with undiminished welfare as compared with present generations. Something like this conception has been advocated by Robert Solow (1974, 1993), who argues that we should aim to maintain sufficient wealth so that future generations will have the opportunity to achieve the same welfare levels as present generations. There are at least two common interpretations of the stable welfare conception: One recommends that we maintain the current level of total welfare; another recommends that we maintain stable average welfare. Solow assumes that welfare or utility will be stable when there is enough capital to support nondecreasing consumption of goods and services and that we should strive to save for the future at a rate that will ensure that the future has as much total capital as we do. If we identify welfare with consumption, as Solow does, then the total view would imply that we should maintain stable rates of consumption over time and across generations. This is in fact what Solow recommends. On the same assumption, the average or per capita view would imply that we should maintain a stable average level of consumption over time. I argue that the total view is absurd and that the average view is seriously incomplete. Neither one should be taken as a guiding ideal to focus environmental policy.

STABLE TOTAL WELFARE

The total view is widely assumed by economists and sometimes defended (Ng 1989). But on reflection it is difficult to accept: Few would celebrate the fact that total consumption remains stable if per capita consumption were to diminish toward zero. But this would happen if population levels increase while the quantity of consumables remains the same; each person's share would diminish even though the total amount consumed remained constant. The view is hardly more acceptable if we drop the identification of welfare with consumption; on any account, stable total welfare is consistent with steadily declining average welfare, so it is possible to maintain stable total welfare levels over time even if each member of each subsequent generation is much worse off than each member of any preceding generation.

STABLE AVERAGE WELFARE

The notion that we should strive to ensure stable per capita or average welfare may be an improvement over the total view because it is impossible to maintain stable average welfare if all members of subsequent generations are progressively worse off. But reflection should make it clear that this conception is also extremely difficult to accept. Average consumption or welfare may remain stable even if more and more people in each succeeding generation are destitute and starving. This could be the case, for example, where the wealth and consumption of a minority increases over time while the majority languishes in poverty and want. Such increasing poverty and inequality reflect the actual state of the world, but few people regard this as a good thing. For these reasons, we should not pursue stable average welfare in our efforts to make our institutions sustainable.

A Defensible Conception of Sustainability?

The Need for an Alternative Conception

My arguments against these alternative conceptions of sustainability have been brief. But perhaps they are sufficient to motivate the idea that we need an alternative conception of sustainability before we can use the concept to frame social or environmental policy. In particular, I emphasize that welfare may be the wrong currency by which to measure sustainability. Economists usually associate welfare with consumption or preference satisfaction, but we cannot easily know what future generations will want to consume, nor can we know what they will prefer. More importantly, our current choices not only may influence the ability of future generations to satisfy their preferences but may sometimes determine which preferences they have. If the satisfaction of some preferences is trivial (such as the desire to watch a mindless TV sitcom), or rep-

rehensible (such as the desire to harm others or engage in gratuitous destruction), then we may have good reason to ignore some preferences or even to
prevent their satisfaction. We may have a current interest not only in seeing to
it that our descendants' preferences will be satisfied but in doing what we can
to ensure that they will acquire certain preferences and not others. Economists
and social scientists often are loath to consider whether different preferences
might be more or less deserving of protection and respect, but such evaluative
nihilism obviously should not guide parents, nor should it guide us in choosing policies that will influence the distant future.

But desires and preferences aside, it is far more important for us to leave
future generations resources to feed themselves and meet their needs than to
ensure that they have adequate access to sitcom reruns on TV. Any theory
that fails to recognize this difference and its significance cannot be a good
theory for evaluating policy. I recommend a dual conception of sustainability involving separate principles for human needs and environmental protection:

Human sustainability: Institutions are humanly sustainable only if their operation does not leave future generations worse equipped to meet their needs
than members of the present generation are to meet their own needs.

Environmental sustainability: Institutions are environmentally sustainable only if
their operation during one generation leaves the earth's environmental systems no more damaged than they were when that generation arrived.

The broader conception of sustainability that I want to defend as a social
ideal combines these two principles. I do not mean to imply that the conception of sustainability embodied by these two principles embodies the
only acceptable normative conception of sustainability: It is a baseline conception that defines what may be only a preliminary goal. If we are currently
unable to meet the needs of many people, then we may have an obligation
to make things better for future generations, not simply to pass on the status
quo. On reflection, we might decide that satisfying needs is too minimal, and
we might discover that it is possible to articulate a conception of sustainability that involves more than need satisfaction. If ameliorating environmental damage is possible and can be achieved without excessive human
cost, it is quite plausible to think that we may have obligations to ameliorate
the damage caused by ourselves and by previous generations so that we leave
things better than we found them. But even if we have positive obligations
to make future generations better off than we are, the obligation not to do
what would make them worse off must be greater. A similar argument can
be made about our obligation not to damage the earth's ecosystems. Because
current institutions are neither humanly nor environmentally sustainable, it
may be appropriate to set our sights on this baseline ideal before pursuing

more ambitious goals. The saying that "the best is the enemy of the good" teaches us that it is often better to define the next positive step instead of outlining a final ideal aim.

Human Sustainability

A principle requiring that present institutions must be humanly sustainable would require that present institutions not reduce the ability of future generations to meet their needs. There are two different ways to reduce the extent of future unmet needs. One is to see to it that the members of future generations will have the ability to satisfy their needs. Another way is to reduce fertility so that future generations will be smaller, so that fewer needy people will come into existence. Empirical studies offer evidence that the former policy serves the latter: By far the most effective way to reduce fertility is to see to it that present people are secure in their ability to satisfy their needs. The relationship between needs, fertility, and population growth is discussed later in this chapter. (See also Wolf 2000; Sen 1996.)

In discussing the stable welfare conception of sustainability, one objection raised was that we cannot know what future generations will want. This fact is less persuasive as an objection to a needs-based conception of sustainability because we can be much more confident about what future generations will need. Standards of need must be culturally sensitive, but when they are appropriately qualified it is reasonable to regard them as cross-culturally and intertemporally valid. Consider, for example, the list of basic needs articulated by David Braybrooke:

The need to have a life-supporting relationships to the environment

The need for food and water

The need to excrete

The need for exercise

The need for periodic rest, including sleep

The need (beyond what is covered under the preceding needs) for whatever is indispensable to preserving the body intact in important respects

The need for companionship

The need for education

The need for social acceptance and recognition

The need for sexual activity

The need to be free from harassment, including not being continually frightened

The need for recreation (Braybrooke 1987, 36–37)

Even if we cannot have certainty about future needs, it is plausible to predict that future generations will need all of these things and that they will be
much worse off if they are arbitrarily deprived of the ability to achieve them.
Surely it is implausible to suggest that we should frame social policy around
the possibility that future generations will be so different from us that they will
not have the needs Braybrooke identifies as basic. Future preferences and wants
may be too obscure as a policy objective, but future needs are not. This provides an important advantage of needs-based as opposed to welfare- or pref-
erence-based conceptions of sustainability.

It is much more difficult to articulate just what we should do, what we
should save, and what policies we should adopt if to ensure that future generations will not be deprived arbitrarily of the ability to meet basic needs. Future
generations may not need petroleum reserves if alternative energy sources are
discovered. They may not need steel or copper if substitutes are developed. It
is an empirical question which policies will best protect the ability of future
generations to meet their needs, and this question cannot be answered by a
normative conception of human sustainability alone. But this empirical question becomes focused once we understand clearly what our aims are. Articulating a conception of sustainability does not answer all of these crucial empirical questions, of course, but it will at least direct us to those that are most
crucial.

Human Sustainability and Intergenerational Justice

It is reasonable to regard human sustainability as a minimal requirement of
intergenerational justice. There are several ways in which such a claim might
be supported. First, we may note that future generations are vulnerable to our
choices and that it is typically regarded as unjust when some people needlessly
deprive others of the ability to meet basic needs. The requirement of human
sustainability is violated only if we leave future generations worse off than we
are ourselves with respect to needs satisfaction, so violating the requirement of
human sustainability would make later generations worse off to benefit previous better-off generations.

But there is an odd feature of such an argument: It treats generations as if
they were individuals. Although earlier generations may be better (or worse)
off than later ones, some members of these same earlier generations may be
much worse off than an average member of the later generation. If the appeal
of human sustainability derives from the high priority this principle assigns to
need satisfaction, then it is plausible to think that present unmet needs are at
least as important morally as future unmet needs. For these reasons, we might
regard it as permissible to address present needs first when we face a tragic
choice between the needs of present and future generations. This minimal pri-

ority for the present is plausible for other reasons as well: We know more about the needs of members of the present generation than about the needs of distant future generations.

A second argument in support of the claim that human sustainability is a minimal requirement of intergenerational justice draws on the ideal contractarian tradition in political philosophy, the most prominent representatives of which are John Rawls and John Harsanyi. Rawls and Harsanyi have both argued that principles of justice are those we would agree upon if we were denied information about our particular station or situation in society. These restrictions on knowledge are necessary, they argue, because such information might lead us to choose rules that would be biased toward our own parochial interests. Rawls calls this mental constraint the veil of ignorance. Parties to the contractual agreement (or choice) behind this veil know facts about society: They know that people have religious beliefs, that they have different conceptions of the good, different wants, and different needs, but they don't know their own condition in life. From behind this veil, not knowing whether one will be a woman or a man, Asian or Caucasian, Muslim or Jew, one will be careful to avoid choosing principles that would arbitrarily disadvantage (or advantage) the members of any such group. From behind the veil of ignorance, one would not choose racist or sexist social principles because one could be disadvantaged by such principles once the veil is lifted. Rawls stipulates that the veil must also deprive individuals of information about which generation they belong to:

> The parties do not know which generation they belong to. . . . They have no way of telling whether it is poor or relatively wealthy, largely agricultural or already industrialized. . . . The veil of ignorance is complete in these respects. Thus the persons in the original position are to ask themselves how much they would be willing to save at each stage of advance on the assumption that all other generations are to save at the same rates. That is, they are to consider their willingness to save at any given phase of civilization with the understanding that the rates they propose are to regulate the whole span of accumulation. . . . Since no one knows to which generation he belongs, the question is viewed from the standpoint of each and a fair accommodation is expressed by the principle adopted. All generations are virtually represented in the original position. (Rawls 1971, 287–288)

Rawls assumes (wrongly perhaps) that the problem of intergenerational justice is solved once we have settled on a fair rate at which present generations should save for the future. But the aim of savings, in Rawls's view, is to guarantee that the members of future generations will be adequately well off and to ensure that earlier generations will not benefit excessively at the expense of later generations.

In his more recent work, Rawls himself articulates a needs principle, stipulating that the goal of meeting basic need is prior to all other principles of justice because, as he argues, these needs must be met if citizens are to be able to understand and exercise the rights and liberties justice is supposed to guarantee (Rawls 1996). He also clarifies the nature of the choice that is to be made from behind the veil of ignorance:

> Since society is a system of cooperation between generations over time, a principle for savings is required. Rather than imagine a (hypothetical and nonhistorical) direct agreement between all generations, the parties can be required to agree to a savings principle subject to the further condition that they must want all previous generations to have followed it. Thus the correct principle is that which the members of any generation (and so all generations) would adopt as the one their generation is to follow and as the principle they would want preceding generations to have followed (and later generations to follow), no matter how far back (or forward) in time. (Rawls 1996, 274)

If we engage in Rawls's thought experiment, imagining that we do not know to which generation we belong, what principle of savings would we hope that our predecessors will have used? From such a position, it is hardly likely that we would choose a principle inconsistent with the requirement of human sustainability because the choice of such a principle would unnecessarily increase the probability that our own needs would be unmet. The case for human sustainability as a requirement of intergenerational justice will be stronger if we accept, with Rawls, that parties to this choice should use maximin reasoning, making the worst possible outcome as good as it can possibly be. Principles of intergenerational distribution that violate the principle of human sustainability will guarantee that some members of some subsequent generations will have their basic needs unmet. And if Rawls is correct to argue that meeting citizens' needs is a first priority of justice, then choosing an intergenerational distribution that is humanly sustainable will be a first priority for a theory of intergenerational justice.

Environmental Sustainability

Not all moral requirements are requirements of justice, and it is not obvious that environmental sustainability can be defended as a requirement of justice. Traditionally, justice has been understood as a relationship that applies between persons, and the obligation to avoid damaging the world's ecosystems is not directly an obligation to others. But because the welfare of future generations and their ability to meet their needs is intimately connected with the health of the world's great ecosystems, environmental sustainability will usually and perhaps always be a necessary condition for human sustainability. This provides an instrumental reason for us to want our institutions and choices to be environmentally sustainable. Beyond this, many of us intrinsically value the integrity of the natural systems of the earth. There is nothing irrational or unreasonable about holding such a value, and it can be (and has been) argued that it is the most appropriate evaluative attitude to take toward natural systems and nature in general. Finally, it may turn out that the best way to prevent environmental damage is to work to make our actions and institutions humanly sustainable. In the next section I argue that there is good reason to think that this is the case.

The definition of environmental sustainability given here relies crucially on a notion of damage. To know what it would be to pass on environmental systems that are no more damaged than the ones we inherited, we need to know what damage is and to have grounds for distinguishing between damage and amelioration. Not all anthropogenic (human-caused) environmental change is damage, and there are hard cases: If the cheetah would have gone extinct anyway, even without human interference and habitat destruction, do we cause damage when we endeavor to prevent or postpone extinction? There have been efforts to reseed stressed coral reefs with concrete "reef-balls," designed to preserve reef habitat as coral dies. If the death of the coral itself is caused by human activity, then that activity surely inflicts damage, but there is some disagreement about whether importing reef-balls themselves constitutes damage or amelioration of damage.

But the existence of hard cases does not undermine the fact that many instances of human environmental damage are clear and uncontroversial. To recognize some human influence as damage, we do not need a full theory of ecosystem health or proper ecosystem functioning. Even if no account of ecosystem health or proper function is acceptable, we may still confidently recognize damage when developers bulldoze roads and condominium foundations in the Sonoran Desert and when logging companies clear-cut ancient forests in Indonesia and Central America. Once again, the articulation of a conception of sustainability will not resolve all thorny policy questions. It will be enough if it sets us working on the right questions. Later in this chapter I consider whether reintroducing wolves serves the aim of human or environmental sustainability. But before considering this question, I discuss one important connection between the two conceptions of sustainability articulated earlier. There is good reason to think that human and environmental sustainability are not only compatible aims, but that they are mutually reinforcing: Often the best way to promote environmental sustainability is to ensure people's long-run ability to meet their needs. And the best way to ensure that people will be able to meet basic needs in the long run often is to minimize and ameliorate anthropogenic environmental damage.

Human Sustainability, Human Needs, and the Growth of Human Population

Human sustainability requires that we avoid imposing on the needs of future generations, at least when this is not necessary to meet the needs of the present. Environmental sustainability requires that we avoid adding to the damage done to the earth's ecosystems by previous generations. But where these values conflict, what are we to do? Some argue that it is inappropriately anthropocentric to recommend that we sacrifice the well-being of the environment for the well-being of people. Others argue that environmentalists are misanthropic because they are willing to sacrifice human needs for the sake of the nonhuman environment. Fortunately, we do not often face a tragic choice in which one basic value must be traded for another. Although it is possible for human and environmental sustainability to diverge so that we will be unable to achieve both, there is reason to believe that such dilemmas are rare in practice. One important reason for this is that poverty, the proliferation of human needs, and rapid human population growth are among the most significant threats to the environment. But evidence shows that the most effective way to reduce human fertility is to address the problem of human poverty and see to it that people's needs are met. Policies that reduce human fertility in this way serve a dual function because fertility reduction would mean that later generations would be less numerous, and meeting their needs would be a less onerous task. And later generations are likely to impose less environmental damage if basic needs are meet and if they are less numerous. (See Sen 1996; Wolf 2000.)

Failure to meet the sustainability conditions identified earlier in this chapter will guarantee human and environmental disaster in the long run. My hope is that a focus on sustainability may direct our attention to the most important of our current environmental problems and provide resources toward their resolution. By directing us toward the most important current issues, this focus may help us more appropriately to set our human and environmental priorities. It is worth noting that sustainability does not seem to be a natural value in one important sense: Looking at the history of the earth, it is clear that many natural environmental processes have been unsustainable. But this is no reason to regard it as an undesirable goal for our own lives and for human institutions in general. The phenomenal economic growth of the past hundred years shows that rapacious and unsustainable treatment of the environment can have spectacular short-term economic benefits. There is no good reason to believe that the global economic market takes the long view or that reliance on markets is consistent with Aldo Leopold's notion that we should think "like a mountain." This puts us in a kind of intergenerational collective action dilemma in which economic advantage to the present may imply grave costs for the distant future. It is much easier to articulate a conception of sustainability than to make our own institutions sustainable. But by keeping objectives clearly in mind, we may better understand the relationship between our policy objectives. Only if we understand our objectives and their relative priority will we be effective in designing appropriate human and environmental policies and in organizing our lives and choices.

Why Reintroduce Wolves?

To this point, much of the discussion here has been highly abstract, but such abstraction is valuable if it can help us to articulate clearly the broader aims for policy and help us to organize present priorities. In this spirit, we may ask what priority we should place on reintroducing rare species when we consider such policies as a means to achieve our broader environmental objectives. In an important sense, wolf reintroduction serves the goal of environmental sustainability defended earlier: The extermination of wolves in the Adirondacks seems an uncontroversial case of environmental damage, and wolf reintroduction seems an appropriate way to alleviate that damage. Such reasoning is implicit in arguments sometimes used to defend reintroduction programs. But different people have very different kinds of reasons behind their various attitudes toward wolf reintroduction. In evaluating the policy, we should examine these reasons carefully. The following list articulates the objectives of many wolf reintroduction advocates. In deciding whether such a policy is appropriate, it is reasonable to begin by considering the terms in which it has been defended:

The objective to forge a sustainable relationship between human institutions and the earth's ecosystems

The objective to promote and preserve natural variety

The objective to protect the earth's ecosystems and avoid gratuitous interference with them

The objective to rectify past environmental wrongs

The objective to protect the interests of human stakeholders

The objective to promote the welfare of wolves themselves and protect them from gratuitous harms

This list surely is not exhaustive; people may have other reasons for advocating wolf reintroduction. But these are surely among the most important objectives advocates seek to promote. To evaluate whether wolf reintroduction is appropriate, we need to consider how well reintroduction serves these objectives and also the relative priority of these objectives. It is worth noticing that these objectives derive, for the most part, from a more basic aim to promote environmental sustainability by ameliorating or minimizing anthropogenic environmental damage. It is difficult to argue that wolf reintroduction serves human needs, but human interests will be considered here as well. In the next section I examine each of the listed objectives in turn. I try to consider whether the objectives are individually appropriate and whether they are well served by the policy of wolf reintroduction.

Wolf Reintroduction: Ordering Our Environmental Priorities

The Objective to Forge a Sustainable Relationship Between Human Institutions and the Earth's Ecosystems

Some people see wolf reintroduction as a way to enact the value of sustainability by repairing damage done by past generations. Most impact on the environment is damaging in one way or another, but wolf reintroduction may be seen by some people as a way to give something back. If we cannot achieve environmental sustainability by avoiding environmental damage altogether, perhaps we can ameliorate the damage we continue to do by repairing the damage done by those responsible for exterminating wolves in the Adirondack Mountains. In the present case, it is environmental sustainability that is most directly served by protecting the Adirondacks and ameliorating human encroachment and damage.

The Objective to Promote and Preserve Natural Variety

The absence of wolves in the Adirondack Mountains means that there is less natural variety than there could be. By reintroducing wolves, we increase local biodiversity and variety. The first two objectives are intimately associated: One might see restoring natural variety as a way to promote environmental sustainability, and wolf reintroduction may serve both aims.

But three major types of policy instrument serve these aims: habitat protection, reintroduction, and legal protection of rare species and individuals (laws such as the Endangered Species Act [ESA]). Clearly these three are not mutually exclusive, and often preservation efforts include all three. But among these three instruments, habitat protection is by far the most effective. Many of us held our breaths in 1995 when the Supreme Court decided *Babbitt v. Sweet Home* (U.S. Supreme Court, no. 94-859), the case that determined that the ESA can be used to protect habitat and not just individuals. If *Babbitt* had been decided differently, the ESA could have been weakened to the point of uselessness. But even as things stand, the ESA protects habitat only incidentally when we can show that endangered individuals live there. A better law would protect habitat itself and would regard the species as important incidentals. For when we protect individual animals we preserve what we know and identify specifically. But when we preserve habitat, this has far-reaching benefits for species we may know nothing about. Clearly the aim to protect biodiversity is served both by reintroduction efforts and by habitat protection. But it is just as clear that these objectives are much better served by habitat protection.

This point is especially crucial in the Adirondack Park, where 60% of the land is privately held and where resident animal populations live in close proximity to human residents. A first priority for environmentalists should be to protect habitat by securing that land from further development. Because reintroduction policies are pursued in the service of an objective to promote sustainability and natural variety, we might do better to channel our energies into habitat protection rather than focusing on wolf reintroduction.

The Objective to Protect the Earth's Ecosystems and Avoid Gratuitous Interference with Them

Although the Adirondacks have been protected from much environmental damage, it is obvious that human activities continue to damage the Adirondack ecosystem. Extirpating wolves was a clear instance of gratuitous interference and damage, so those who accept that human beings have an obligation to minimize and mitigate such damage should also recognize that this objective provides potential support for the policy of wolf reintroduction.

The Objective to Rectify Past Environmental Wrongs

The objective to minimize interference is closely allied to the objective to ameliorate past environmental damage and rectify the wrongs human activity has inflicted. The extermination of Adirondack wolves resulted directly from human mistakes and human malice toward wolves, who have long been vilified and regarded as "varmints" to be killed. A policy of extermination was active until the 1890s, when the last known Adirondack wolf was shot. Many of us accept the obligation to avoid unnecessary interference with the world's remaining "intact" ecosystems and to do what we can to mitigate the effects of past interference. Wolves belong in the Adirondacks, or so it is said. And we are at fault for the fact that they're not there anymore. Many of us have a sense that we share responsibility for these policies and for the wrongs that were perpetrated under them. Although we were not among those who actually shot the Adirondack wolves, we are culturally and socially related to those who did. This sense of responsibility may lead us to hope that we can in some measure mend the damage done by our ancestors and avoid, as much as possible, inflicting more damage of our own. Wolf reintroduction may be seen as a way to rectify wrongs perpetrated by members of our group and for which we bear, or feel we bear, a measure of responsibility.

The Objective to Protect the Interests of Human Stakeholders

This brings us to the thorny issue of human interests in the Adirondacks and in the reintroduction of wolves. Human stakeholders who have interests that may be promoted or thwarted by a policy of wolf reintroduction include Adirondack residents, visitors, hunters, recreationists, and people who value the Adirondacks from a distance. It is impossible to articulate a single, coherent set of human interests in wolf reintroduction: Different people have widely different interests, and no policy can please everyone. Many Adirondack residents are staunchly opposed to wolf reintroduction. They express a wide variety of concerns: Some worry that wolves will prey on pets and livestock, that wolves may unacceptably reduce the number of deer and other prey species, and that wolves will make the region more dangerous. Others strongly support the policy of reintroduction, either because they see wolf reintroduction as serving other objectives or because they regard it as intrinsically desirable.

If we consider the reasons that lead people to support or oppose wolf reintroduction, we find that there is confusion and misinformation on both sides of the debate: Many who oppose reintroduction do so because of their false beliefs about the danger posed by wolves. Many are still in the grip of a mythical stereotype of wolves as vicious and bloodthirsty. Supporters of wolf reintroduction often articulate an opposing mythical image of wolves as noble, heroic, and poetic. As always, it is inappropriate to burden other species with the virtues and vices we find in our own behavior: Wolves are neither noble nor vicious, neither heroic nor villainous. We can admire and respect them or hate and vilify them, but clearly our attitudes reflect more about us than about the wolves.

As far as possible, our attitudes toward wolf reintroduction should be based on understanding, not myth. Better understanding of wolves is sure to promote the eventual extinction of the traditional myths that wolves are evil and vicious. And when people's expressed interests are obviously based on misleading stereotype, it is tempting to suppose that their true interests are different from those they express. If so, this would not justify overriding expressed interests, but in such circumstances it might be appropriate to fight the misleading myths with education. For the time being, present human interests do not speak strongly for or against wolf reintroduction.

But human stakeholders also include future generations who will inherit the Adirondacks as we leave it and who should not forgive us if we fail to leave it intact. Our interest in future generations is twofold at least: On one hand, we want to leave them opportunities to live good and satisfying lives. On the other hand, we may also have a keen interest in shaping their preferences and values. Many of us share a present desire to foster our children's interest and respect for wilderness and for wolves, to give to our children our own love for the wilderness. Perhaps this cannot be done without bringing them to the wilderness. We don't just care that they should have opportunities to get what they want; we want them to have opportunities to nurture the right kinds of wants and values. It is plausible that our ability to provide such opportunities is promoted by the existence of wilderness areas that have been spared damage from human encroachment and have been restored, as much as possible, where damage has been done. It is plausible to think that restoring wolves to the Adirondacks would serve this interest.

The Objective to Promote the Welfare of Wolves Themselves and Protect Them from Gratuitous Harms

Those who work to reintroduce wolves and those who support their efforts often are driven by an intense love and respect for wolves themselves. It is arguable that this love and respect are founded on understanding rather than myth because these people work intimately with wolves, have studied them, and have thought deeply about them. But those who love and respect wolves must feel the pull of conflicting aims when considering reintroduction. Certainly reintroduction programs reduce the danger that wolves might go extinct, but wolves are not in serious danger of extinction. And policies that promote the perpetuation of a species may still be contrary to the interests of every member of that species. Some might recommend that we secure the future of wolves by promoting a market for wolf pelts and stuffed wolf trophies. Herefords and hogs are not in danger of extinction, and some people have seriously suggested that the best way to protect endangered species is to find commercial uses for them so that entrepreneurs will have reasons to domesticate them. But many of us would vigorously oppose "preservation" by domestication. Our interest in protecting the future of the wolf and other endangered species is far more complex than simply an interest in preserving global genetic diversity. Such diversity is an important aim, but this aim must be balanced against others that conflict with it.

We can and must distinguish the interests of wolves from the "interests" of "The Wolf": the species Canis lupus. The fact that they are not the same sometimes creates conflict between those interested in protecting animal welfare and those interested in protecting biodiversity. Those who admire and respect the wolf must acknowledge a strong presumption against subjecting wolves to the kind of treatment they receive during reintroduction. In the case of the wolves that might be introduced in New York State, these animals will suffer either captive breeding or being trapped and taken from their more familiar environment. These intelligent social animals may be separated from lifelong companions and friends, although this may be avoided if (as in Yellowstone) groups are taken in pack units. They will be set loose in an unfamiliar place full of dangers and face local hostility from a significant and armed minority of the local human inhabitants. We would not wish such a fate for a friend, nor should we be quick to wish it for a spectacular and intelligent animal for whom we have respect. The record for reintroduced wolves still is a mixed bag. Some populations have done well and have produced pups. But in other cases transplanted wolves have been disoriented and harmed by the reintroduction process and have been unable to survive (see Ferguson 1996). Experience is valuable, and those responsible for wolf reintroductions have learned from past mistakes, but from the perspective of an individual wolf or pack it is probably better to avoid capture and relocation even in the best circumstances.

But this may not always be the case. Transplanted wolves often are taken from areas where they already face serious dangers. Wolves in British Columbia and Alaska are already hunted and are sometimes killed in systematic wolf reduction policies. These policies are subject to severe criticism, and one response might be to work to protect wolves without transplanting them. But if wolves face the alternative between being shot in British Columbia or transplanted to the Adirondacks, then surely it is better for them to be transplanted.

Conclusions

The objectives I have considered here do not all point toward the same conclusion, and thoughtful people will come to different conclusions when they evaluate the wolf reintroduction proposal against these objectives. In particular, many people will agree that our legitimate objectives to protect the earth's ecosystems, avoid gratuitous interference with them, and rectify or ameliorate past environmental wrongs speak powerfully in favor of wolf reintroduction. Other objectives considered here offer somewhat weaker support for the policy because human interests and the interests of wolves are ambiguously served by wolf reintroduction and because wolf reintroduction is not an effective way to promote the sustainability of human relationship to the environment.

Considered from a broad perspective, the value of wolf reintroduction may be more symbolic than real. If our main objective is to prevent current environmental damage and to mitigate past environmental destruction, then it may be a mistake to focus on gains that are primarily symbolic because this may distract us from more pressing environmental and social concerns. In the present context, I believe that such distraction is especially costly because our current situation is serious: We are in the middle of a genuine environmental crisis of unprecedented significance. In a crisis, it is especially important that we keep our priorities clearly in view.

If we are in a crisis, what kind of crisis? Currently, its most threatening face is that of habitat destruction. The present rate at which we are consuming and destroying irreplaceable forests, deserts, coral reefs, rivers, and oceans is not just alarming; it should be terrifying to anyone who is aware of it. There are estimates that fully 10% of the earth's coral reefs have already been destroyed, mostly in the last 20 years. And it has been argued that current rates of change will cause another 60% to disappear in the next 20 to 40 years. E. O. Wilson predicts a 20% decrease in world biodiversity in the next 20 years or so, and some sober ecologists regard this estimate as conservative (see Eldredge, Chapter Twenty, this volume). At present, we seem to be consuming the environment on which we depend and on which future generations will rely. It should be obvious that the status quo is not sustainable in any important sense.

Alarming statistics such as these might leave us bewildered and at a loss: What can we do in the face of a global problem? In our desire to do something, we settle for symbolic gains and feel that we've done our bit. But there are concrete local steps we can take to respond even to the global crisis, and among the most effective steps are those that move wildlands off the development schedule. Working to reintroduce wolves may make us feel better and may give us the impression that we're doing something practical to respond to the current environmental crisis. We may feel that we're giving something back, mitigating past wrongs. But even if this sense of responsibility is laudable, it does not justify us in making inappropriate triage decisions. In the current context, spending resources on wolf reintroduction is a bit like setting a broken arm when others around us suffer from uncontrolled arterial bleeding. It's a good thing to do, but let's secure the habitat first. Perhaps we can begin by protecting that privately owned 60% of the Adirondack wilderness. This could be accomplished by increasing the proportion of protected public lands within the Adirondack Park and further restricting development on privately held land.

At present, the best way to remedy environmental damage and protect the Adirondacks may be to increase environmental protection and public ownership of land in the Adirondacks. But this must be done in a way that is responsive to the interests of current Adirondack residents, many of whom have expressed deep opposition to plans to increase public holdings. Local opposition to public land acquisition is almost as strong as opposition to wolf reintroduction. This presents a serious obstacle to Adirondack protection, but perhaps it is an obstacle that can be overcome once residents realize that increased public ownership is likely to serve their economic interests. Privately held plots of land will become far more valuable as private holdings become more scarce and as more and more land is protected. Increased environmental regulation will be costly for current residents, but can continue to offer benefits through the distant future. Currently the highest priority for those who love the Adirondacks should be to find mutually beneficial arrangements that will increase current environmental protections while promoting the interests of residents and stakeholders. Because I am convinced that environmental protection really is in the long-term interests of most people, I believe that such arrangements can be found. Later, once the Adirondacks are restored and better protected, it will be both easier and more appropriate to turn our attention toward the reintroduction of native species such as the wolf.

I cannot claim to have discovered a uniquely appropriate method for weighing the considerations I have raised. The evaluation of any significant policy always is complicated by many different factors, and where decisions are complicated reasonable people will disagree. But I argue that at present the best way to serve the goals of sustainability, natural variety, and the health of the world's ecosystems would be to spend our resources and energies promoting habitat protection and making human needs secure. Many of us would like for there to be wolves in the Adirondacks. But in the current crisis, focus on wolf reintroduction may cause us to neglect more pressing environmental concerns.

Acknowledgment

I would like to thank Victoria Davion, Jennifer Everett, Cecilia Herles, Jake Howard, Mary Midgley, Bryan Norton, Ernest Partridge, and Rebecca Sharp, who have discussed many of these issues with me and whose ideas have influenced this paper. I benefited from discussion of this paper at a conference titled Wolves and Human Communities, held at the Museum of Natural History in New York City in October 1998, and from Virginia Ashby Sharpe, who provided excellent comments on the entire manuscript.

Part VII

Consciousness: Ecological and Cultural

The book's final part situates the question of wolf reintroduction in the contexts of ecology and evolutionary biology and urges a broadened cultural consciousness of our role in ecosystem change.

Chapter Nineteen, "The Ecological Implications of Wolf Restoration: Contemporary Ecological Principles and Linkages with Social Processes," is by Steward Pickett and Ricardo Rozzi. Pickett is senior scientist at the Institute of Ecosystem Studies in Milbrook, New York. Rozzi is a Chilean ecologist and environmental ethicist as well as cofounder of the Institute of Ecological Research, Chiloe, Chile.

Lest we fall back on an ill informed or romanticized view of the "balance of nature," Pickett and Rozzi explain what, from an ecological perspective, is a healthy, intact, or complete ecosystem. Such systems, they say, are those that are capable of persisting and adjusting in the face of changing environments. This notion of completeness therefore emphasizes a dynamic understanding. The primacy of such dynamics is illustrated by the persistence and adjustment characterizing one of the most fundamental processes in the natural world: evolution.

Continuation from existing lineages and adjustment are the hallmarks of evolution. Variation, or flexibility in changing environments, emerges from mutation and genetic mixing. These genetic changes are the raw material of organisms' ability to respond to different environments. If a range of environments is present or if the environment changes from generation to generation, then the offspring that match the environment will tend to persist and reproduce, whereas those that do not match or that match less well probably will die or do poorly.

This process generates and maintains the natural biological diversity of the earth. But biodiversity is not an end product; it is part of an ongoing process. Evolution and the production of biodiversity have weathered catastrophes of nearly global proportions and filled the most amazing array of environments

255

with life. Indeed, biodiversity itself has become a fertile matrix that stimulates the evolution of additional types of organisms. Each kind of physical environment and each kind of organism presents new challenges and hence new evolutionary opportunities. The process of evolution can be likened to a cycle of challenge and opportunity. There is no final goal, only a continuing process of creativity and winnowing, flexibility, and adjustment.

The same sort of process, they observe, characterizes the ecological world. Ecology is the science that studies the interactions affecting organisms and the systems that contain organisms. Although at any one time it may seem as though ecologists are studying static things, in reality these things are the temporary reflections of ongoing processes of interaction. When ecologists study the forest cover of northern New York, for example, they must understand how the glaciers affected the land and set the stage for the forest to be established over 12,000 years since the glaciers retreated. They must understand the past episodic events, including natural ones such as windstorms and human-caused ones such as logging and fires in logging slash, that determine the current forest species composition and physical structure. They must understand how interactions with animals that consume plant leaves and seeds, and the interactions with fungi and bacteria in the soil, affect the forest. There are myriad interactions, but perhaps the most important thing about those interactions and the systems they define is that they are hardly ever constant. Each of these changes, short or long, frequent or infrequent, creates a crisis for some species while it creates opportunities for others.

There are essentially no ecosystems on earth where human influence is absent. The biological world, when lightly touched by humans, functions by carrying out interactions over large distances. Organisms move to avoid stresses or to take advantage of new sites opened by natural disturbances. An important feature of the world with light human effect is that there are spatially isolated areas where new evolutionary diversity can be generated. Where human influence on ecosystems is more extreme, however, the ability of the biological world to recover from change and crisis is reduced.

Restoration, say Pickett and Rozzi, is one way in which we might positively contribute to the capacity of ecological systems to recover their ability to create variety and adjust to changing conditions induced by large human impacts. A number of hurdles face ecological restoration, however, all of which are grounded in ignorance of how the natural world works. These include the tendency of humans to focus on the short term rather than the long term, our tendency to invest restoration with a "nature knows best" attitude, the failure of macroeconomics to account for the fact that ecological processes are the very foundation of economies, and the ways in which human structures fragment natural landscapes.

Management, they argue, as a social process with an ecological core, must compensate for these mismatches if we are to maintain the goods and services as well as the ethical connections we have with the land.

The volume's final chapter, "A Hard Sell: The Cultural-Ecological Context of Reintroducing Wolves to the Adirondacks" is by Niles Eldredge, research paleontologist at the American Museum of Natural History. Eldredge argues that the political and logistical difficulties of wolf reintroduction to the Adirondacks are exacerbated by a more general resistance to acknowledging the present crisis in biodiversity, namely, that we are in the midst of a sixth global extinction event, with potentially grave consequences for our own species, *Homo sapiens.*

One of the reasons why it is so difficult for people to grasp the significance of the current mass extinction—of approximately 30,000 species a year—has to do with a critical step in human cultural and ecological history, a step that profoundly turned us away from the natural biological world. That step, 10,000 years ago, was the invention of agriculture, which allowed humans for the first time to transcend, and indeed abolish, their ecological niche.

The invention of agriculture, the literal taking control of production of the greater part of the food resources consumed by the community, meant stepping away from, even outside of, the local ecosystem. Indeed, he says, to plow a field and plant the seeds for one, two, or at most three desired food crops is to declare the territory off limits to all other species: plants, of course, but animals, fungi, and all elements of the microbial world unable to withstand the changes wrought by humans as they plow, fertilize, plant, harvest, and reset the fields for subsequent planting.

There were many profound consequences to the origin of agriculture. Agriculture meant settled existence, which further meant the growth of cities, economic division of labor unlike anything seen in the past, and social stratification. Of all of these, the most profound was population growth beyond the carrying capacity of local systems.

Population numbers, Eldredge observes, are regulated by the productivity of the local ecosystem. The carrying capacity of the local ecosystem for most bands of human hunter–gatherers (i.e., known from the few that survived to the present in postagricultural times) usually is around 40 people, although the number may be as great as 70. Thus the total number of individuals of any species living on earth is the number of individuals within each local population times the number of different populations of that species living in different ecosystems. The numbers typically are statistically stable over long periods of time.

With the invention of agriculture, and with people no longer living inside local ecosystems, there was no longer an upper limit on the number of peoples who could live in any one setting. This meant that the traditional cap on the total human population—the cap that limits the numbers of all other species, past and present, living on earth—suddenly was removed. And this is the simple reason why the human population has mushroomed—slowly at first, but logarithmically spiraling out of control as we enter the new millennium—from that mere 5 million 10,000 years ago, to the 6 billion we have today.

The most dramatic effect of our ability to live outside of local ecosystems is that we have created the illusion that we no longer are a part of nature at all, nor do we need healthy, vibrant ecosystems to survive.

However, Eldredge says, human impact on the environment, especially noticeable over the past 500 to 1,000 years, signals our true position vis-à-vis the natural world: We have become a global species internally connected not just by our sharing of genes but also by our economic behavior.

If *Homo sapiens* has indeed become a concerted economic force as a species, it stands to reason that there must be an economic system in which we operate. That system, Eldredge says, is the entire biosphere, the summation of all the world's ecosystems that forms the system in which we are playing a direct, concerted role.

In other words, we have not left nature with the invention of agriculture and the abandonment of local ecosystems. Instead we have redefined our position in the natural world. To the extent that we understand this, we recognize that it makes a very great difference indeed whether we continue to degrade ecosystems and drive species to extinction.

These paleontological reflections urge patience in any deliberative process regarding wolf reintroduction. The year-round residents of the Adirondacks have legitimate local concerns, as do all local peoples faced with outside intervention in their affairs, however well intentioned that intervention may be. But they are also like every one else in the world, raised in the belief that the natural world matters to us only in the sense that we can take whatever we think that we need.

Human ecological concerns, whether within ecosystems or in the more recent agricultural phase, have never involved much thinking ahead beyond storing grains and other provisions for the dead of winter. In our preoccupation with the short term, we have little or no track record of thinking ahead, even for our children's entire lifetimes after we are gone.

Efforts such as the present plan for wolf restoration in the Adirondacks, though promising some short-term rewards, are really about planning for the longer-term future. Perhaps, says Eldredge, that is the most radical part of the plan.

As this last part illustrates, the question of wolf reintroduction is a question about the intricacy and interdependence of human and natural systems, systems that are connected by time and space and disconnected by complex forces both outside of and within our control. The daunting reality of our responsibility for managing earth's ecosystems in a way that is sustainable for long-term human well-being and ecosystem health should encourage us to forge connections where we can. Efforts to establish trust between the human players in these decisions is a worthy step in this process.

Chapter Nineteen

The Ecological Implications of Wolf Restoration: Contemporary Ecological Principles and Linkages with Social Processes

Steward T. A. Pickett and Ricardo Rozzi

The human dimensions of wolf restoration range widely across the social, political, and ethical realms. The intersection of wolf restoration with each of these realms points out important ecological principles that can help inform the social dialogue that is the heart of any successful restoration effort. Here we explore what restoration accomplishes in ecological terms and relate the ecological processes that are the target of restoration to constraints on their successful reestablishment. The constraints point to the key ecological ingredients in successful restoration and expose the ecological justification for the responsible management that must accompany restoration in a human-dominated landscape.

What Does Restoration Accomplish?

In discussing the potential for reintroducing the wolf to areas where it has been extirpated, "healthy," "intact," or "complete" ecosystems often are mentioned. These terms are useful in a social dialogue, but they must be further specified if they are to accurately reflect rigorous ecological principles (Haskell et al. 1992).

From the perspective of contemporary ecological understanding, the terms *healthy* and *intact* imply that "complete" systems are those that are capable of persisting and adjusting in the face of changing environments (Grumbine 1994). Therefore, this notion of completeness emphasizes a dynamic under-standing. The primacy of such dynamics is illustrated by the persistence and adjustment characterizing one of the most fundamental processes in the natural world: evolution. Continuation from existing lineages and adjustment are the hallmarks of evolution (Gould 1977). Variation, or flexibility in changing environments, emerges from mutation and genetic mixing. These genetic changes are the raw material of organisms' ability to respond to different environments. If a range of environments is present, or if the environment changes from generation to generation, then the offspring that match the environment will tend to persist and reproduce, whereas those that do not match or that match less well probably will die or do poorly.

This process generates and maintains the natural biological diversity of the earth. But biodiversity is not an end product; it is part of an ongoing process. Evolution and the production of biodiversity have weathered catastrophes of global proportions and filled the most amazing array of environments with life. Indeed, biodiversity itself has become a fertile matrix that stimulates the evolution of additional types of organisms (Huston 1993). Each kind of physical environment and each kind of organism presents new challenges and hence new evolutionary opportunities. Evolution can be likened to a cycle of challenge and opportunity. There is no final goal, only a continuing process of creativity and winnowing, flexibility, and adjustment.

We introduce this contemporary view of evolution because the same sort of process characterizes the ecological world. Ecology is the science that studies the interactions affecting organisms and the systems that contain organisms. Although at any one time it may seem as though ecologists are studying static things, in reality these things are the temporary reflections of ongoing processes of interaction. As Aldo Leopold noted in 1949,

> The image commonly employed in conservation education is "the balance of nature." For reasons too lengthy to detail here, this figure of speech fails to describe accurately what little we know about the land mechanism. A much truer image is the one employed in ecology: the biotic pyramid. (Leopold 1949, 251)

> In the beginning the pyramid of life was low and squat; the food chains short and simple. Evolution has added layer after layer, link after link. Man is one of thousands of accretions to the height and complexity of the pyramid. (Leopold 1949, 253)

When ecologists study the present distribution and abundance of a species, they really must understand the genetic potential of the species, the evolutionary characteristics it has accumulated, the patterns of migration and local extinction it may have experienced, the important interactions it has with prey, predators, or beneficial neighbors, and so on. Similarly, when ecologists study the forest cover of northern New York, for example, they must understand how the glaciers affected the land and set the stage for the forest to be established over 12,000 years since the glaciers retreated. They must understand the past episodic events, including natural ones such as windstorms and human-caused ones such as logging and fires in logging slash, that determine the current forest species composition and physical structure. They must understand how interactions with animals that consume plant leaves and seeds, and the interactions with fungi and bacteria in the soil, affect the forest. There are myriad interactions, but perhaps the most important thing about those interactions and the systems they define is that they are hardly ever constant. The interactions fluctuate with yearly changes in weather and with long-term shifts in climate. They respond to the occasions of extreme drought that appear on the scale of decades or to the huge storms that occur perhaps only once a century or over even longer time periods.

Each of these changes, short or long, frequent or infrequent, creates a crisis for some species while it creates opportunities for others. Windstorms that blow down some old trees create new sites in which sun-loving seedlings can do well. The conversion of a living tree to a rotting log robs a specialized leaf-feeding insect of its food while creating a moist habitat for salamanders and forming a raised platform on which slow-growing seedlings of hemlock can survive. This generation of variety in the environment becomes the pattern to which ecological processes respond (Kolasa and Pickett 1991). However, variety in the environment is not constant. There are times and places in which it is higher than others. These patterns over space and time are not constant. Therefore, organisms find an unpredictable template to which to respond. Unpredictability lends additional opportunity, if it is not too extreme, for the flexibility and variety of organisms to flourish.

Over the long term, species adjust to the shifting and accumulating variety in the environment, including other species. Therefore, ecological understanding entails a long-term view (Likens 1989). Like the social view, originating with the Onondaga, that asks what the effect of any action today is on the seventh generation hence (Thomas, Chapter Five, this volume), the ecological view involves assessing the flexibility and adjustment in the biological components of systems over the long term. The history of life as reflected in evolution is one such long view. So are the ecological processes that result in the adjustment within ecosystems. The natural portion of the earth is engaged in an ongoing cycle of flexibility and adjustment. There is no necessary end point, and there is no need to seek an external or final goal. Adjustments, including those within ecological systems to other internal components or species, are the name of the game.

How Have Flexibility and Adjustment Worked Without Humans?

Humans have become such a commanding presence in relation to ecological processes that it may seem futile to ask how the world works in their absence or where their presence and influence is low. Indeed, today there are essentially no places where human influence is absent (McDonnell and Pickett 1993). Even the remote lakes of the Adirondacks are affected by deposition of acids derived from distant pollution sources. Likewise, the forests there are in the path of the introduced beech bark disease, and it seems unlikely that they will escape its effects. However, asking how the world works when humans are a local component of ecosystems and landscapes, rather than a regional or global dominant, serves as a useful reference to understand how ecosystems work and to expand our understanding of what restoration seeks.

The biological world, when touched lightly by humans, functions by carrying out interactions over large distances. Organisms move to avoid stresses or take advantage of new sites opened by natural disturbances. Hot spots for resources that organisms need may move throughout a landscape. For example, as beavers abandon dams, marshes are created where locally high productivity can occur. Or the lightgaps in forest open suddenly when trees fall in a storm but are closed and therefore become less desirable for species needing high levels of resource availability as neighboring or infilling trees grow over a few years or decades.

An important feature of the world with light human effect is that there are spatially isolated areas where new evolutionary diversity can be generated. A partial or low flow of genes from large, central populations to the nooks and crannies of a landscape make it possible for local adjustments to be enshrined in the genetic codes of those isolated members of a species. Sometimes those isolates become fully differentiated species, and in other instances they share some genetic exchange with the rest of the species. Such nooks for novelty have been important ingredients in the capacity of wild nature to adjust to the huge variety of environmental exigencies that occur.

In addition to needing unencumbered spaces over which to operate, the processes of the natural world (such as plate tectonics and geologic and climatic variation) operate over long time periods. Of the important processes affecting ecological systems, perhaps the one most people assume to take a long time is evolution. Although important mutations or new genetic combinations can arise very quickly, sorting out the wealth of variation may take much longer. It typically takes new characteristics of organisms a long time to spread through a population or a species. It may also take a long time for other features of the organisms to adjust effectively to the new characteristic. In other cases, where characteristics are not of crucial importance or cost to the organisms involved, there may be a slow drift in the proportion of that character compared to its alternatives (Futuyma 1986).

One of the processes that take time is the renewal of biological diversity after a large extinction episode. Huge losses of biological diversity have occurred throughout the geologic history of the earth. Those natural periods of extinction were caused by episodes of global scope such as an immense asteroid strike or the drifting together of continents so that habitat for near-shore marine organisms was greatly reduced. Although the species richness of the earth was drastically reduced by these events, after each one the remaining organisms took advantage of the new environments and the presumed access to greater amounts of uncontested resources, and diversity slowly began to flourish again (Wilson 1992). The diversification of life on earth has increased despite these great episodes of extinction because there was time for the remaining organisms to adjust. In addition, recalling that the availability of vast space permits organisms to migrate, find special and different habitats, and survive local or even regional catastrophes, it is clear that the long times and large spaces were available in the world before human domination. Unfortunately, the widespread toxification and transformation of the landscape happening today reduce the ability of the biological world to recover after extinctions, whether they are natural or human caused (Pickett et al. 1992).

Although many ecological processes occur over shorter time spans than the grand trajectory of evolution, ecological processes still take time. For instance, when a forest burns or a flood wipes out an ecosystem that had been present in the bottomlands of a river, surviving plant species or seeds usually sprout quickly or invade, or animals move back in from refuges nearby. But the full occupancy of the site may take years or decades. Similarly, the ecosystem functions, such as nutrient flow and storage of organic matter in the soil, may take decades or centuries to recover to levels that existed before the disturbance. These examples of ecological succession, in which communities and ecosystems respond to natural and some human-made disturbances, commonly are quite slow, and legacies of past disturbances are present in many ecosystems. For example, in New England and adjacent New York, the glacial ice cover began to retire from its maximum southern limit about 17,000 years ago and was followed by recolonization of spruce, pine, beech, hemlock, and oak (Gaudreau 1988). About 10,000 years ago, hunting-and-gathering peoples began to proliferate, and horticultural practices and use of fire became increasingly common, especially after 1,200 years ago (Mulholland 1988). With the arrival of European settlers, the diversity of animal species and the forest cover declined dramatically and persistently until the nineteenth century, when the forest covered less than 30% of its presettlement area. During the twentieth century, a marked decrease in agriculture and pasture land has been accompanied by a recovery of the forest, reaching 80% of its original area. Deer, beaver, and coyote have concomitantly increased (Foster and Motzkin 1998). This extremely dynamic history of natural and human disturbances in New England and New York emphasizes the significance of an evolutionary–ecological approach, rather than an emphasis on fixed scenarios (Pickett and Parker 1994), for restoration projects such as the Adirondack wolf reintroduction.

These examples of the nature of evolutionary and ecological change can be summarized by saying that the natural world functions as a result of the creation of new opportunities, the generation of variety in organisms and ecological systems, and the assortment of different kinds of organisms and the systems to which they contribute. Nature is an ongoing cycle of variation and adjustment to new variation. It is a self-organizing process that continues because of the great space over which it can range and the long times it can take to respond. People draw off some of the capital and richness of nature to build, power, and inspire themselves and their societies. Ecological services are derived from this flow of variation and assortment that constitutes the business of nature.

The Gap Between Ecological Processes and the Social Environment for Restoration

If restoration aims to permit the capacity of ecological systems and the ecological components of our human-dominated world to recover their ability to create variety and adjust to changing conditions, how can it be accomplished? What constraints on restoration does the case of the wolf call to our attention?

Perhaps the most serious problem facing ecological restoration is the common experience of people and groups. The scope of individual memory is on the order of one generation, and the extent of personal experience is likewise constrained in space. The size of humans and the range of environments they can personally experience during an ordinary life limit what we know of the natural world. As a result, many processes of the natural world that are necessary for continued variation and adjustment are invisible to individual humans. This invokes Leopold's (1949, 141) conclusion that "the hidden meaning in the howl of the wolf, [is] long known among mountains, but is seldom perceived among men." Such invisibility might be called the limit of common sense. So much of crucial importance to the natural world's ability to continue to respond flexibly and to provide the ecological goods and services on which we all ultimately depend never appears on the radar screen of common sense. Common sense is good for dealing with the local contingencies of life under normal circumstances, but it blinds us to the deep machinery of the world, which needs vast length and scope.

Another incongruity is that between ecological processes and some cultural assumptions we make about nature. As Dizard (Chapter Eight, this volume) and Midgley (Chapter Fourteen, this volume) point out, whatever else the natural world is, it is also a canvas on which people and societies paint their values. Within that big space, different peoples, religions, cultures, businesses, and so on, have painted meanings that range over great contrasts. Some believe nature to be evil and to need suppression; others believe it to be fragile and worthy of protection. One extreme among these views is that of eighteenth-century British philosopher David Hume (1986, 179), who said,

> When we recommend an animal or a plant as useful and *beneficial,* we give it an applause and recommendation suited to its nature. As, on the other hand, reflection of the baneful influence of these inferior beings always inspires in us with sentiments of *aversion*. The eye is pleased with the prospect of corn-fields and loaded vineyards; horses grazing, and flocks pasturing: but flies the view of briars and brambles, affording shelter to *wolves* and serpents. (emphasis added)

New England colonizers, steeped in British philosophy, promoted livestock and other "beneficial" domestic animals, whereas they attempted to eliminate predators. William Cronon (1983, 132) states in his book about the ecological history of New England,

> Few things irritated colonists more than finding valuable animals killed by "such ravenous cruell creatures." The Massachusetts Court in 1645 complained of "the great losse and damage" suffered by the colony because wolves killed "so great nombers of our cattle," and expressed frustration that the predators had not yet been successfully destroyed. Such complaints persisted in newly settled areas throughout the colonial period.

Such a strong aversion to wolves by European settlers was widespread (Kellert et al. 1996). In 1669, the Virginia Burgesses enacted a law requiring the Indians in the colony to kill wolves as a yearly tribute to the villagers (Gohdes 1967). Similar actions led to the disappearance of wolves in New England, despite their abundance in this region before the eighteenth century (Lohr and Ballard 1996). The last wolf was killed in Connecticut in 1837, in New Hampshire in 1887 (Merchant 1989). In Massachusetts, Thoreau (1856/1984, 220) wrote in his journal, "I spend considerable of my time observing the habits of the wild animals, *my brute neighbors.* . . . But when I consider that the nobler animals have been exterminated here—the cougar, panther, lynx, wolverine, *wolf,* bear, moose, deer, the beaver the turkey, etc., etc.—I cannot but feel as if I lived in a tamed, and, as it were, emasculated country" (emphasis added). Thoreau performs a decisive turn from the settlers' attitudes previously described. Thoreau calls the wolf and other "nobler animals" his "brute neighbors," a denomination that departs from British utilitarianism and brings him nearer to his other neighbors, the American Indians.

Among the Indians dwelling in the forests of the Northeast, Great Lakes, and Northwest, kinship and reciprocity between human beings and animals

was a widespread belief. Clans were based on legends of descent from ancestors such as the wolf and used their images in their works of art (Hughes
1996). The Kwakiutl of the Pacific Northwest practiced a wolf-inspired
method of hunting, the wolf way to drive deer, and hunting was seen as a holy
occupation involving exchanges of gifts (Hughes 1996; Overholt and Callicott
1982). Chipewyan, isolated from European influence in the northern forests,
retained until the nineteenth century a universally observed rule of not killing
wolves or wolverines (Martin 1984). Ojibwa in the Great Lakes told in their
narratives that prey animals, such as beavers, are willing to give themselves to
humans for food. To do so, beavers and other prey animals require that humans
live up to a certain set of strict obligations involving the making of offerings
and the proper treatment of the bones of the dead prey (Overholt and Callicott 1982). This sophisticated and intricate web of ancestral, indigenous beliefs
and environmental practices, conveying respect for wolves and prey, was soon
noted by naturalists and defenders of wilderness such as Henry David
Thoreau, John Wesley Powell, and John Muir. The latter left us an expressive
illustration of the Indian conservation attitudes toward wolves. In his diary,
Muir (quoted by Hughes 1996, 138) wrote about a conversation he had one
night during a canoe trip in Alaska,

> I greatly enjoyed the Indians' camp-fire talk this evening on the
> ancient customs, how they were taught by their parents where the
> whites came among them, their religion, ideas connected with the
> next world, the stars, plants, the behavior and language of animals
> under different circumstances, manner of getting a living, etc.
> When our talk was interrupted by the howling of a wolf of the
> opposite side of the strait, Kadachan puzzled the Minister with the
> question, "Have wolves souls?" The Indians believe that they are
> wise creatures who know how to catch seals and salmon by swim-
> ming slyly upon them with their heads hidden in a mouthful of
> grass, hunt deer in company, and always bring forth their young at
> the same and most favorable time of the year. I inquire how it was
> that with enemies so wise and powerful the deer were not all
> killed. Kadachan replied that wolves knew better than to kill them
> all, and thus cut off their most important food supply.

The acknowledgment of such diverse and contrasting views of nature and
attitudes toward wolves among philosophers, New England colonizers, North
American Indians, and American environmentalists conveys a key lesson for
ecological restoration. Human attitudes toward nature, and hence human
impacts on the environment, are not intrinsically negative. Among historical
periods and geographic regions, we discover a tremendous diversity of relationships between humans and nature, within our Western tradition as well as

among indigenous groups, that can convey both negative and positive ecological effects. This recognition clarifies the scene and gives hope for restoration efforts because conflicts do not occur between human society and nature as universal and abstract entities but as particular cultural forms of social groups and their environments (Rozzi et al. 1997).

Within the range of values that people project on the natural component of the world is a problematic assumption. Implicit in the discussion of restoration is the idea that "nature knows best." If something seems amiss, the "balance of nature" will reassert itself. Or perhaps people look to apparently uninhabited regions that seem to bear little trace of humans and find that the balance they perceive there will serve as a useful model for their own approach to the natural world. But this takes nature to be essentially static or completely resilient. Note that all of these assumptions about how the natural world is are very different from the contemporary ecological understanding of nature as a process of variation and adjustment. To the extent that we use any of these cultural assumptions to guide restoration or the subsequent management (or lack thereof) of natural phenomena, we are as likely to do harm as to do good. Fundamental to successful restoration, which may be motivated by cultural values and common sense, is the appreciation of the way the world really works.

Another troubling mismatch between ecological processes and social and cultural institutions is the failure of macroeconomics to account for how the natural world works. The deepest flaw here is the failure to recognize that economies and societies depend on the products of natural variation and adjustment (Norton, Chapter Seventeen, this volume). In other words, ecological processes are the very foundation of economies and societies. Many specific assumptions of mainstream macroeconomics clash with ecological reality. Included is the failure of prices to capture ecological processes. In addition, discounting environmental damage into the future means that economic costs and benefits will not come into equilibrium, and environmental effects probably will be invisible. Furthermore, the common indices of macroeconomics, such as the gross national product, do not account for environmental damage (Norton 1991b). Economics ultimately fails to account for the space and time needed by the natural processes on which people depend.

Many aspects of the law are similarly incongruent with how we understand the world to work (Sax, Chapter Ten, this volume). The law is an ancient and evolving institution, but much of the legal structure under which we operate does not recognize that the natural world is intertwined with the social and political worlds. To the extent that legal systems reflect any understanding of the natural world, it is an ancient but incorrect one. If cultural assumptions such as "the balance of nature" or "the great chain of being" in which the natural world is subordinated to human concerns inform our laws or judicial decisions, then the ability of the natural world to vary and adjust continuously may be compromised. If legal decisions do not consider the needs of the natural world, then how can they comment effectively on restoration and management? A specific feature of the law that fails to appreciate natural processes is the structure of property regimes that ignore the public interest in the provision of ecosystem services by private land (Lehmann 1995). In addition, there is no acknowledgment of the value of ecological services that may be associated with parcels of land, except in the most obvious cases of nuisances produced. Furthermore, the law operates in a piecemeal fashion and does not address the interconnected nature of the natural processes it intentionally or accidentally regulates (Caldwell and Shrader-Frechette 1993). Finally, the adversarial system may increase the likelihood that natural processes will be set against immediate, private human desires. Given that natural objects and processes, such as species, rivers, and evolution, do not have legal standing, they have long been the loser in these contests.

Journalism is one of the most visible social phenomena that fails to appreciate ecological processes. The ability of the media to report the current scientific understandings of the natural world is seriously compromised by the adversarial approach the media take to issues they report. For instance, in the presumed interest of "fairness," reports of scientific issues almost always include a dissenting opinion. But for many questions that scientists have had time to work on and sort out, the scientific conclusions have reached the status of a consensus. The fact that some dissenters can almost always be found does not necessarily mean that the consensus is threatened. Of course, sometimes the scientific community does radically change its conclusions. It is during such periods of upheaval and discovery of new phenomena or new approaches to well-known phenomena that contrary opinions reflect how the scientific process works. These distinctions rarely are made by the media, which follow a model of error-and-blame that seems to work out something like this: Identify a new event, discover what is damaged or who is hurt or benefited, and identify who is responsible. This kind of model, exemplified in the case of reporting fires in nature, does a disservice to the understanding of ecological processes. Wildfires are reported using an urban fire model (Smith 1992). Built structures are not supposed to burn. Therefore, when there is a fire in a settlement, some error or crime has been committed. Who is responsible, and who has lost life or property as a result of the error? Although wildfires often damage people and property, there are other important aspects of fires in nature. In fact, whether set by people or not, they may be a part of the larger and longer pattern of natural disturbances to which an ecological system has adjusted. The fires in Yellowstone National Park in 1988 were part of a longer pattern of drought-maintained severe fires that had occurred on a scale of centuries in the region (Christensen et al. 1989). Many components of the natural system had features that allowed them to deal effectively with such fires. In fact, in the absence of fires of that sort, or management events that can mimic the key effects of those fires, the system would lose some of the features for which we value it. Furthermore, in the Yellowstone region, Indians burned areas periodically and extensively until they were removed from the park in the 1870s. Park rangers and managers suppressed fire as much as they could. After a century of such suppression, the result was "huge expanses of over-aged lodgepole pine forests, full of dead wood and other hot-burning fuels" (Norton 1991b). Thus, the fires of 1988 reflect a complex of factors. These include a long-term, natural drought cycle, elimination of the traditional Native American fire regime, and suppression of fires by park managers. Certainly the exclusion of wolves played a role in the vegetation–fire relations in the park. Wolves affect the browsing intensity by mammalian herbivores. Therefore, they are another component in the mix of factors controlling the vegetation. For example, the present aspen stands in the northern range of the park were established under a unique combination of past factors (Romme et al. 1995). The period during which the aspen stands were established a century ago followed a major fire, was wet, and had an abundance of browsers. In addition, wolves, which prey on the herbivorous elk, were present (Romme et al. 1995). This complex interplay between large-scale disturbances, indirect effects between predators such as wolves and vegetation, and historical relations between Indians and the Yellowstone ecosystem were poorly addressed by journalists focused on the "sudden" sensational fire. The mismatch between journalism and the understanding of ecological processes can be understood as a mismatch of temporal scale: The media require that their reports be of breaking news, and most ecological processes are slow and ongoing. Thus, many of the most important features of the natural world are invisible to those who report the news and consequently to those who depend on the news for their understanding of the world.

The largest mismatch between the way the world works and the state of the human-dominated sphere is the very structure of human-dominated landscapes. If natural processes need large space and long times to proceed unencumbered, then the landscapes we have built over millennia of agriculture and centuries of industrialization are inimical to the unfettered phenomena of nature (Forman 1995). Our contemporary landscapes are fragmented, leaving few but highly isolated places where natural ecological processes and evolution can operate as they did for eons. Species are threatened because they have little space to support their activities or because their pathways for migration in the face of environmental change are blocked. Any given ecosystem operates on local inputs of solar energy and material resources present within its boundaries or transported by wind, water, or organisms. Humans subsidize their activities in local systems by adding energy from fossil fuels and resources from elsewhere. Often fossil fuels, material resources, and capital are transported over large distances from their point of origin to their point of use. Such subsidies mean that ecological systems today are much less governed by local or regional processes. A telling example of this subsidy is the spread of invasive exotic species, which threaten local ecosystems' ability to produce or maintain variety that is suitable to the exigencies typical of that locality (Putz 1997). Likewise, new stresses, some that exceed the capacities of any organism to respond evolutionarily or even adjust physiologically, are more widespread over human-dominated landscapes than in the past. It is this highly fragmented and stressed landscape in which restoration is necessary but highly constrained (Saunders et al. 1991). How can we restore systems and key species such as the wolf on such a limited stage?

Management and Restoration in a Fragmented World

Fragmentation can be understood as a complex of constraints. It suggests the spatial simplification of the landscapes, the associated severe stresses that have become endemic and constant, and the new and severe disturbances that humans practice and promulgate. To compensate for this new, constrained reality, society must act in well-informed ways to maintain or restore the fundamental ecological processes that generate flexibility and variety in systems and allow subsequent adjustment to new opportunities. Society must also compensate for the telescoping of time over which systems must respond. Management may be able in many cases to compensate for the short times available for ecological adjustment.

However, to practice the kind of informed management needed to compensate for the shredded and limited landscapes of the current human-dominated era, some of the social limits detailed in the last section must be overcome. Perhaps most confining is the cultural assumption that "nature knows best" (Cronon 1995b). Perhaps this is a useful metaphor when thinking about the world lightly or only locally touched by humans, but nature's knowledge is constrained in its application in fragmented contemporary landscapes. We must erase some of the layers of interpretive paint we have slathered on the canvas of nature so that we can understand the ecological processes of flexibility and adjustment. Then we may be able to re-create at least some of their key effects in our occupied and constrained landscapes. To assume that we must keep hands off the natural world as it exists in most places today means that we ignore the other hands that have constrained nature in the past or those that constrain it from a distance (Russell 1993). The hand of society already is pervasive in nature. As French philosopher Larrère (1996, 117) suggests in her essay "Ethics, Politics, Science, and the Environment: Concerning the Natural Contract," "there is no pristine nature; there is only 'hybrid' nature that has been shaped by both natural and cultural forces." Recognizing ecological processes needed in restoration means that we can apply society's hands in an informed way through management. Management proved crucial in minimizing uncontrolled hunting and trapping of wolves within and outside of protected areas. Likewise, management that is socially and ecologically informed is playing a major role in the destiny of reintroduction and management efforts in Ontario (Algonquin Park), Central Idaho, and Yellowstone National Park (Forbes and Theberge 1996; Fritts et al. 1997).

It is important to remember that management is a process by which an ecological problem is identified, possible responses that can maintain or restore missing ecological processes are articulated, decisions are made on which (if any) action to take, and results are monitored and assessed for their ecological and social effects (Holling 1978). Note that management doesn't always mean taking action in the natural world. Managers do not always have to do something.

Management does require knowledge, openness to learning new things about the system and new approaches to management, and the capacity to adjust the management to accommodate new circumstances that appear in or affect the system (Christensen et al. 1996). Too many management schemes in the United States have been fixed and inflexible or not monitored, compromising their ecological effectiveness and social value.

One important lesson to be learned from the prospect of wolf restoration in the Adirondacks is the role of people and social processes in the decision. Ecologists have come to recognize that it is unproductive and unrealistic to think of natural systems and human systems as separate. They are part of a larger, more inclusive system (Burch 1988). That larger system depends on the fundamental ecological processes for its continuation, but it also is governed by social and institutional structures and processes and feedbacks between social and natural processes. Decisions about whether and how to restore the wolf or any other biological entity or ecological process must incorporate the social components of the system as well. How will local and regional economies be affected, how will social institutions respond and adapt, and how will the human population change in size or structure as a result? These are just some of the questions that sustainable ecological management must address. Including the social dimensions and the local people in the decision about wolves in the Adirondacks is laudable, necessary, and consonant with a growing integration of social dimensions with ecological processes (Burch et al. 1997).

Thinking Like a Mountain Means Acting Like a System

Aldo Leopold's (1949) classic essay "Thinking Like a Mountain" shows the maturity of his thinking about the relationship of people, wolves, and nature.

The ethical dimension of his essay calls us to put people and the land, or ecological systems more broadly speaking, together when we make social and individual decisions that affect the natural world. His awareness of the fierce green fire in the dying wolf's eyes personalizes that connection and calls attention to a key species in ecosystem processes. The metaphor of the mountain calls humanity to take a long-term view, to honor the longevity of the mountain, and to see a wide view from the mountaintop. In this way the image shows the concern with the long times and large spaces that ecological flexibility and adjustability have taken throughout most of the planet's history. That insight of contemporary ecology, which was not yet fully formed in 1944 when Leopold penned that compelling metaphor, is one we can now use as a scientific tool in service of the ethical view that arose from the spark between Aldo Leopold and the dying wolf.

Thinking like a mountain means taking responsibility for the mismatches between the fundamental and necessary natural processes and the "unecological" constraints of culture, economics, common sense, law and media, and the fractured and stressed landscapes that exist today. Management, a social process with an ecological core, must compensate for these mismatches if we are to maintain the goods and services as well as the ethical connections we have with the land. The wolf, with its cultural meanings and its complex and generalized role in ecosystems and landscapes, is an ideal scruple to stimulate our thinking about the nature of ecological processes and their relationship to a healthy society.

Chapter Twenty

A Hard Sell: The Cultural–Ecological Context of Reintroducing Wolves to the Adirondacks

Niles Eldredge

The proposed reintroduction of *Canis lupus* to the Adirondack Mountain region of northern New York State is an exciting initiative in conservation biology. Biologists and conservationists have come to realize two important lessons as they have sought ways and means to stop, or at least slow down, the tremendous wave of extinction that is claiming some 30,000 species a year (Wilson 1992). First, recognizing that human-caused habitat destruction and ecosystem degradation are overwhelmingly the major cause of species loss, we have learned that the top priority in conservation efforts must be to identify and preserve tracts of largely intact habitats, preferably linked to other such tracts through corridors. Second, we have learned that, resilient though nature is, the direct, proactive management or stewardship of ecosystems also is necessary, at least in most cases, for ecosystem restoration and persistence to become a reality.

The proposed wolf restoration project in the Adirondacks fits these twin criteria nicely. The 6-million-acre Adirondack Park is an amalgam of publicly and privately held land, zoned and classified from frankly commercial plots to vast tracts considered "forever wild." The setting thus in place, the next question is how much and what kind of human input is appropriate to reconstituting forest, wetland, and alpine ecosystems in the Adirondacks, systems that developed after the glacial retreat some 10,000 years ago, persisting up to the advent of European colonization, when farming, hunting, lumbering, and tourism began to take their toll.

True, species removed from ecosystems often find their way back in the normal sequence of ecological succession. Surviving populations in adjacent regions send out colonists as a matter of course. The recent advent of coyotes *(Canis latrans)* back in many regions of the Northeast, including the Adirondacks, is a graphic case in point. Moose, too, are making a slow comeback, wandering in from neighboring regions to resume residence in the watery habitats from which hunters drove them in the 1880s.

But other species that were there before European colonization are not likely to come back to the Adirondacks on their own, and *Canis lupus* is an excellent example of such a species. I will leave it to others to assess the ecological impact that reintroduction of wolves is likely to have on other species and, of course, on human economic activities. It seems to me that the purely biological consequences are not at all straightforward. For example, in 1996 a small pack of coyotes took down a deer about one-half mile from our summer camp in the Adirondacks. If it is true (as I have heard in passing) that the coyotes thriving in the Adirondacks are somewhat larger than their predecessors were, it is possible that the economic role (niche) of coyotes in the park overlaps with the prospective role that reintroduced *Canis lupus* would be expected to play. The standard expectation would be for the wolves to supplant the coyotes altogether or at least outcompete them for the larger game species, especially Virginia white-tailed deer. Indeed, that would be one of the many interesting biological hypotheses that the "experiment" of wolf reintroduction would allow us to test.

Many people, including me, will enthusiastically embrace the proposed reintroduction of wolves to the Adirondacks. They will do so for a host of reasons, including love of wolves, desire to see precolonial nature restored, and sheer intellectual curiosity. But many others will look upon this project with dismay, and tempting though it may be for some advocates of this project to dismiss such resistance as a reflection of simple ignorance, I believe it imperative for the prime movers of any restoration project—especially involving such a notorious (rightly or wrongly) species as *Canis lupus*—to come to grips with not only the fact of opposition but also with the reasons why such opposition is typically so deeply entrenched.

Three Sources of Opposition

I see three basic sources of opposition to the wolf restoration project. The first, fascinating and perhaps obvious as it may be, is the deeply ingrained fear of and hatred for wolves in Western cultural history. The second, easily confounded with the first, actually constitutes the third great lesson conservation biology has or should by now have learned from experience: No conservation project involving large tracts of land can be successful in the long term unless the economic needs, practices, and concerns of the local citizenry are taken explicitly into account and incorporated into the plan. Fortresslike walled-off enclaves replete with "keep out" signs are doomed to failure, especially in places such as the Adirondacks, with its quiltlike pattern of private and public land ownership.

Again, speaking anecdotally but based on an aggregate 40 years' experience as an outsider, transient visitor, and lover of the Adirondacks, I see the full-time residents of the Adirondacks as utterly typical of peoples the world over who are faced with a difficult economic setting and initiatives generated by outsiders (though often with local sympathy) to conserve the biodiversity of their local area. By no means do I mean to imply that all full-time Adirondack inhabitants are opposed to wolf reintroduction; I know from my own experience that that is not true. But resistance to wolf reintroduction is not based simply on ignorance and historical cultural prejudice against wolves, including the largely imagined worries about wolf depredation on livestock and the baseless worries about human safety.

Rather, much of the opposition reflects the simple fact that several of the counties within the Adirondack Park are among the poorest of New York State; indeed, much of the populace of some of these counties is forced to go on welfare in the tough winter as the economy shrinks. People everywhere feel that it is their right to decide what to do with their land, and especially in economically deprived regions, those rights are typically seen to include harvesting natural resources (especially timber in the Adirondacks) and otherwise doing whatever it takes on the land to eke out a living. I am heartened that these issues are addressed by a number of contributors to this volume, but before moving on to the third source of opposition, I must emphasize that wolf restoration in the Adirondacks will not succeed in the long term unless the economic fears posed by wolves are allayed, and especially unless a substantial segment of the Adirondack human community can be both included in the project and shown how their own lives stand to be enriched—monetarily, and perhaps even esthetically and spiritually—by the reintroduction of wolves into their region.

The third source of opposition to efforts in biological conservation is even more subtle than the challenges posed by prejudice (i.e., specifically against wolves) or alienation on economic grounds of local peoples. In a nutshell, biological conservation remains a hard sell; many people simply don't get it. When informèd that we are in the midst of the greatest mass extinction event since the demise of the dinosaurs and so many other plants, animals, fungi, and microorganisms 65 million years ago, most people simply shrug and think, "Who cares? What does it matter to me?" To those of us actively concerned about the biodiversity crisis we now face, the real question becomes not "Why should I care?" but rather "Why don't most people seem to care?"

We at the American Museum of Natural History have just completed a new exhibition devoted to biodiversity and the rapidly mounting loss of functional ecosystems and species on the planet. The 11,000-square-foot Hall of Biodiversity was conceived as the museum's first "issues" hall, the first exhibition that transcends the depiction of the natural world as pristine and actively confronts human destruction of habitat and roles in the current Sixth Extinction. In mounting this exhibition, we knew that most people are simply unaware of what the term *biodiversity* means and unaware of the mounting loss of species and the threat to their lives (and to humanity in general) posed by the current biodiversity crisis.

Biodiversity and Human Evolution

We endeavored to create a hall of great beauty where the vast panoply of life could be viewed in a sweep of the eye and the visitor could think, "Life is amazingly rich and beautiful; do you mean all this is under threat?" The hall asks the visitor to contemplate four simple questions: What is biodiversity? Why should we care about it? What is happening to destroy biodiversity? And what can we do to stem the tide of the Sixth Extinction?

The answers are kept simple: Biodiversity is the sum total of all species living in all the world's ecosystems (see Norton, Chapter Seventeen, this volume). The hall is laid out to mirror the dual nature of biodiversity: Evolutionary biodiversity is the vast spectrum of perhaps 10 to 13 million species, everything ranging from bacteria, eukaryotic microorganisms, plants, fungi, and animals. All of these species, from the smallest and simplest to the largest and most complex, are descended from a single common ancestor that lived more than 3.5 billion years ago. We have arranged a smorgasbord of evolutionary diversity along the Wall of Life, the only place in the world where a museum visitor can take in, again at a glance, the enormous variety of living things.

The world's species are the actors in the drama of life, which is played out in the world's ecosystems. Ecological biodiversity, the other face of biodiversity, consists of the interlocking array of the world's ecosystems covering the surface of the earth. In the Hall of Biodiversity, we explore one such ecosystem in detail (in a modernized version of the American Museum's time-honored tradition of minutely accurate diorama construction) in a walk-through experience replete with sounds, smells, and high-definition background films depicting a segment of rainforest in the Central African Republic. The forest is shown as intact and pristine at one end and very much damaged by human activity (in this case, slash-and-burn agriculture) at the other end.

Throughout the hall, the values of biodiversity are brought out with clarity and simplicity: the fact that humans use more than 40,000 wild species of plants, animals, fungi, and microbes, both directly for food, shelter, and clothing and indirectly, in the technological arena, as the genetics of wild species are explored and extracted for medical and agricultural use. We speak of the enormous importance of intact ecosystems for regulating critical chemical cycles (such as carbon and nitrogen), producing atmospheric free oxygen, maintaining stable climate regimes, and cycling (by now desperately depleted) fresh water for human consumption.

We explore the fact that humans are mimicking the comets that nearly destroyed life 65 million years ago by converting lands for agricultural use and the subsequent growth of cities, overexploiting biological resources (most notably, perhaps, the world's fisheries and forests), and introducing alien species.

The exhibition does not shy away from the root cause of all this destruction: the extraordinary growth of human population, a veritable explosion that has seen our numbers rise from an estimated 5 million (the average estimate; see Cohen 1995) to a whopping 6 billion in just 10,000 years (the date, coincidentally, of the founding of the modern Adirondack environment). Six billion people, coupled with the uneven distribution and especially consumption of the world's resources (both biological and otherwise), is the underlying cause of the Sixth Extinction. Our explorations of what might be done to stem the tide of the biodiversity crisis understandably speaks of stabilizing human population, developing sustainable economic practices, and using modern conservation techniques. The evidence is compelling that we have been successful in getting these messages across to our museum visitors, who number around 3 million annually.

Yet herein lies a further tale, one that takes us closer to the heart of the mystery of why people do not immediately, automatically, and intrinsically see why they should care about ecosystem destruction and species extinction. It turns out that the very root of the problem—the human population explosion, with its attendant disparities in resource consumption—itself hinges on a critical step in human cultural and ecological history, a step that profoundly turned us away from the natural biological world. How we came to degrade ecosystems and destroy species hinges on the same sequences of events that also made us not notice (or care, if we did notice) what we were doing to the earth and the life it supports.

Elsewhere (Eldredge 1995, 1998) I have claimed that human evolution can be read in ecological terms as a process of culture (defined here for convenience simply as traditions of learned behavior and practice) gaining ascendancy over purely biological adaptations over a period of several million years. To illustrate the point, I like to think of the African complex of waterways, riverine forests, and open savannas still preserved in special places, such as Botswana's Okavango Delta (Eldredge 1998). Next to the Adirondacks, the Okavango is my very favorite complex of ecosystems for a host of reasons, but especially because it vividly calls to mind (because it so closely resembles) the ecosystems of 2 to 3 million years ago in east Africa's Rift Valley System, as preserved so beautifully in famous places such as Tanzania's Olduvai Gorge. That this, one of the few stands of Eden left, is every bit as threatened as any other ecosystem in the world is especially poignant, given the Okavango's status as an almost unique remnant of our own birthplace.

Human evolution, like that of all other species, was not a slow, steady eventless slog from ape through ape-man up to our present selves, the picture of gradual evolution handed down to us from the nineteenth century. Rather, the evolutionary history of all of life is much more a matter of stability of species and ecosystems until the inevitable physical disruption that sends ecosystems into disarray and drives species extinct. The vast majority of evolutionary change comes in brief spurts as new species rapidly evolve and ecosystems are rebuilt after the extinction phase. This is true of the five global mass extinction events, but it is equally true of myriad more regional, less all-encompassing extinction events that nonetheless have triggered the bulk of life's evolution. Thus we can focus on a few well-defined, critical events that have shaped human evolutionary history to capture the essence of the human ecological–evolutionary story.

Such an event occurred in Africa between 2.8 and 2.5 million years ago. As detailed especially in the work of Yale University paleontologist Elisabeth S. Vrba, the earth's climate cooled between 10 and 15 degrees centigrade during that 300,000-year interval. The wet woodlands that predominated in eastern and southern Africa absorbed the gradual cooling and drying of the ambient atmosphere for awhile. But then a threshold was reached, and the wetter woodlands rapidly gave way to more open grasslands, with riverine forests and copses of remnant woodland dotting what had suddenly become a much more Serengeti-like open sea of grasslands.

Some species, ecological generalists such as impalas that are adapted equally well to woodlands as open savannas, survived unscathed and unchanged. But many species of animals and plants simply disappeared. Either they recognized suitable habitat and thus survived elsewhere, or they simply became extinct. And then, in a matter of thousands of years, new species began to appear. Some trickled in from elsewhere (as moose and coyotes have been doing in the Adirondacks), already adapted to grassland habitats. But others evolved right there in a reaction to the fragmentation of habitats that is traditionally (and correctly) considered a source of the isolation needed to drive a reproductive rift between populations of an ancestral species.

Among the extinct species and among the newly evolved were members of our own hominid lineage. Just about 2.5 million years ago, *Australopithecus africanus,* a 4-foot-tall, 100-pound (i.e., males; females were smaller), upright bipedal hominid with a brain about the size of a chimp's (about 400 cubic centimeters; for comparison, our brains are roughly 1,400 cubic centimeters) became extinct. Itself something of an ecological generalist, this species probably scavenged animal kills and collected nuts, tubers, seeds, and berries; it may have occasionally hunted small game.

Replacing that species almost immediately were two other hominids: the first member of the robust australopithecines of the genus *Paranthropus* and the earliest member of our own genus, the species *Homo habilis*. The paran-thropines, superficially gorillalike, were massively built ecological specialists; they concentrated exclusively on eating gritty tubers and other vegetation. Typical of ecological specialists, the line rapidly diversified into at least five dif-ferent species, dying out completely after a brief evolutionary sojourn of only a million years.

The other species, *Homo habilis*, fared differently; with its much bigger brain (about 750 cubic centimeters), this was the first hominid species we know that definitely used tools. Crude as they were, their Olduwan choppers are thrilling to contemplate; they are the first definitive evidence of material culture and hence of cultural traditions used to inform the ecological exis-tence of one of our ancestral species.

The next big event in human ecological–evolutionary history is possibly the last time one can point to a correlation between an environmental event and a corresponding (and presumably reactive) event in our evolution. This occurred around 1.65 million years ago, when another cooling snap gripped the globe and when the first of the four great glacial advances southward from the north pole in both Eurasia and North America began.

The evolutionary event in question was the appearance of the earliest members of the *Homo erectus* lineage, once again in Africa. These people stood nearly at the height of modern humans, differing from us most markedly only in terms of their brain sizes which, at about 1,000 cubic centimeters, were advanced but not to the degree reached in later phases of human evolution. Their tools were advanced over those of *Homo habilis* and consisted of a vari-ety of axes, choppers, and scrapers. Recently discovered evidence suggests that they had mastered the use of fire.

One of the most remarkable events in the entire history of life occurred at the onset of the second glacial, just under 1 million years ago. Remains of some early hominids have been discovered in Eurasia that predate this event, but nothing like the rapid proliferation of *Homo erectus* almost throughout the Old World about 0.9 million years ago. This is remarkable: A tropically adapted species expanded its range precisely as the second glacial epoch began. It is dif-ficult to demonstrate any species expanding its ecological niche at any time in the past 3.5 billion years; species expand their geographic ranges but always by tracking familiar habitat. Yet here was a species that went from the warm trop-ics into the frozen north, defying the ages-old response of all species to move toward the equator during glacial advances and becoming extinct if suitable, familiar habitat could not be found. It seems to me that fire, plus the more sophisticated tool kit and undoubtedly other aspects of material culture (cloth-ing would have been called for), underlay this radical ecological shift.

The third phase of human ecological–evolutionary history worth noting in this connection is the African exodus of our own species, a movement that began perhaps 100,000 years ago, some 50,000 or 100,000 years after we evolved there. With our species, we note great ecological as well as purely geographic expansion. We were still living as hunter–gatherers, playing direct roles inside local ecosystems. But wherever we entered ecosystems not only new to our own species but never occupied by our ancestors, the wildlife did not recognize us as the predators we are. The result: We disrupted ecosystems and drove many species extinct, mostly if not entirely by simple overhunting. And most anthropologists now agree that among the very first of the species that fell to human-induced extinction was our collateral kin, the Neanderthal peoples who had evolved as a separate species in Ice Age Europe and western Asia. Thus, whenever modern humans showed up, we touched off a wave of extinction: 40,000 years ago in Australia, 12,000 years ago in North America, 8,000 years ago in the Caribbean Islands, and 2,000 years ago in Madagascar. We had become extremely efficient hunters, and the naive faunas that we encountered were simply no match for us.

Thus cultural traditions, including efficient material cultural items, informed the human niche increasingly through the last 3 million years or so. But although the range of environments in which the hunter–gathering mode of life was played out, always within the context of local ecosystems, nothing truly presaged a true cultural and ecological (but not an evolutionary) revolution: the invention of agriculture, beginning approximately 10,000 years ago, perhaps at first in the Near East but followed swiftly by independent invention in a number of other areas.

Ecologists and historians often cite the invention of agriculture as a major change in the human ecological niche. It was far more than that: In effect, it was the abolition of the human ecological niche in the sense, once again, that niches are the characteristic roles that members of a species living in populations play in the context of their local ecosystems. The invention of agriculture, the literal taking control of production of the greater part of the food resources consumed by the community, meant stepping away from, even outside of, the local ecosystem. Indeed, to plow a field and plant the seeds for one, two, or at most three desired food crops is to declare the territory off limits to all other species: plants, of course, but animals, fungi, and all elements of the microbial world unable to withstand the changes wrought by humans as they plow, fertilize, plant, harvest, and reset the fields for subsequent plantings.

Preagricultural peoples were hunter–gatherers, living in the 3.5-billion-year tradition of playing an active, concerted role inside local ecosystems. Postagricultural people, in stark contrast, became the first members of any species in the entire history of life not to live as small populations inside local ecosystems. Little wonder that we saw ourselves as different from the "beasts of the field," as recounted so tellingly in that famous passage in Genesis. The early agriculturalists no doubt realized they were some form of animal life, but they also saw that they were a very different kind of animal life from anything else they could point to. And, certainly in an ecological sense, they were right.

Of course, there were many profound consequences to the origin of agriculture. Agriculture meant settled existence, which further meant the growth of cities, economic division of labor unlike anything seen in the past, and social stratification. Everything from the loftiest cultural achievements to the vilest social ills flowed from this ecological change. But of all of these, the one closest to the present theme is simply this: Population numbers are regulated by the productivity of the local ecosystem. The carrying capacity of the local ecosystem for most bands of human hunter–gatherers (i.e., known from the few that survived to the present in postagricultural times) usually is around 40 people, although the number may be as high as 70. Thus the total number of individuals of any species living on earth is the number of individuals within each local population times the number of different populations of that species living in different ecosystems. The numbers typically are statistically stable over long periods of time.

With the invention of agriculture, and with people no longer living inside local ecosystems, there was no longer an upper limit on the number of peoples who could live in any one local setting. This means that the traditional Malthusian cap on the total human population—the cap that limits the numbers of all other species, past and present, living on earth—suddenly was removed. And this is the simple reason why our numbers have mushroomed—slowly at first but logarithmically spiraling out of control as we enter the new millennium—from that mere 5 million 10,000 years ago to the 6 billion we have today. And despite recent encouraging signs, that number may double again by the mid–twenty-first century.

Why don't we care that we are causing the demise of some 30,000 species a year? Simple: We correctly perceive that we stepped outside of the natural rule of ecosystems when we invented agriculture. At the same time, that invention removed the Malthusian cap, human population has exploded, and as a consequence, and however inadvertently, we are degrading ecosystems and driving species to extinction at a rate not seen for the past 65 million years. Hence my comment at the outset of this chapter: The very factors that are causing us to destroy ecosystems and eliminate species at the same time blind us to what we are doing.

But there is yet one more event in human cultural–ecological history to recount, and this one makes all the difference as we begin to look around at what we are doing to the planet and its ecosystems and species. Human impact on the environment, especially noticeable over the past 500 to 1,000 years, signals yet another change in our position vis-à-vis the natural world: We have become a global species. But we have become a very special kind of global species, different from other global species (most of which became global by tagging along with humans, such as the fruitfly *Drosophila melanogaster*). We are an internally connected species, not just in terms of our sharing of genes in common (all species have that property by the very definition of the term *species*) but also in terms of our economic behavior. Paul Kennedy (1993) tells us that human beings exchange $1 trillion in goods and services globally every day.

If *Homo sapiens* has indeed become a concerted economic force as a species, it stands to reason that there must be an economic system in which we operate. What is that system? It seems to me it is the entire biosphere, the summation of all the world's ecosystems (sometimes called Gaia) that forms the system in which we are playing a direct, concerted role: We extract resources from the earth, continue to use its living species, and, of course, have a heavily detrimental effect, as if we did not realize that we are actually a part of this larger system.

In other words, we have not left nature with the invention of agriculture and the abandonment of local ecosystems. Instead, we have redefined our position in the natural world. We have to come to understand that and, as a result, to see that it makes a very great difference whether we continue to degrade ecosystems and drive species extinct.

And so, to return the project of wolf reintroduction to the Adirondacks, I can only urge patience in the very critical task of weaving in the concerns of year-round Adirondack residents into the picture. They have legitimate local concerns, as do all local peoples faced with outside intervention in their affairs, however well intentioned that intervention may be. But they are also like every one else in the world, raised in the belief, which used to be true, that the natural world matters to us only in the sense that we can take whatever we otherwise self-sufficient human beings think we need. We need to be aware of this 10,000-year entrenched attitude and work to change it to a realization that every corner of the earth, however degraded or nearly pristine it might be, needs our attention and our proactive stance.

Human ecological concerns, whether within ecosystems or in the more recent agricultural phase, have never involved much thinking ahead beyond storing grains and other provisions for the dead of winter. In our preoccupation with the short term, we have little or no track record of thinking ahead for even our children's entire lifetimes after we are gone. Efforts such as the present plan for wolf restoration in the Adirondacks, though promising some short-term rewards, are really about planning for the longer-term future. Perhaps that is the most radical part of the plan.

References

Adirondack Park Agency. *State of New York Adirondack Park State Land Master Plan.* New York: Ray Brook, 1979.

Alworth, S. "Gentle predators." *Daily Hampshire Gazette,* February 24, 1997, 8.

Anonymous. "Vaguely worded ads force Alaska to delay wolf kill." *Star Tribune,* December 22, 1992, 20A.

Armstrong, K. "An estimate of the Adirondack white-tailed deer population." *NYS DEC Region 5 Report,* 1998.

Askins, R. Letter to the editor. *Jackson Hole News,* December 22, 1993, A5.

Austin, J. L. *How to Do Things with Words.* New York: Oxford University Press, 1962.

Babbitt v. Sweet Home, 515 U.S. 687 (1995).

Babcock, H. "Protection of wildlife and wildlife habitat: a background principle of state property law." In Georgetown University Law Center, Continuing Legal Education. *Successfully Litigating Regulatory Takings Claims,* September 24–25, 1998.

Baker's Ex'rs v. Kilgore, 145 U.S. 487 (1892).

Baldwin, A. D., Jr., Judith DeLuce, and Carl Pletsch, eds. *Beyond Preservation: Restoring and Inventing Landscapes.* Minneapolis: University of Minnesota Press, 1994.

Ballard, W. B., L. A. Ayres, P. R. Krausman, D. J. Reed, and S. G. Fancy. "Ecology of wolves in relation to migratory caribou herd in northwest Alaska." *Wildlife Monographs* 135, 1997.

Ballard, W. B., R. Farnell, and R. O. Stephenson. "Long distance movement by gray wolves." *Canadian Field Naturalist* 97 (1983): 333.

Barnard, E. *In a Wild Place.* Boston: Massachusetts Audubon Society, 1998.

Barrett v. State, 116 N.E. 99 (N.Y. Court of Appeals, 1917).

Barry, B. *Democracy, Power, and Justice.* Oxford, UK: Clarendon, 1989.

Bartlett, J., comp. *Familiar Quotations: A Collection of Passages, Phrases, and Proverbs Traced to Their Sources in Ancient and Modern Literature.* Ed. Emily Morison Beck. 14th ed. Boston: Little, Brown, 1968.

Beckerman, W. "'Sustainable development': is it a useful concept?" *Environmental Values* 3 (1994): 191–209.

Bellah, R. *Habits of the Heart.* Berkeley: University of California Press, 1985.

Berg, K., L. Hatfield, and J. Brave Heart. *Gray Wolf Coalition Statement on Status and Delisting of the Wolf.* Minneapolis: Gray Wolf Coalition, 1998.

Berg, W. and S. Benson. *Updated Wolf Population Estimate for Minnesota, 1997–98.* Grand Rapids: Minnesota Department of Natural Resources, 1999.

Berry, T. *The Dream of the Earth.* San Francisco: Sierra Club Books, 1988.

Bishop, R. "Endangered species and uncertainty: the economics of a safe minimum standard." *American Journal of Agricultural Economics* 60 (1978): 10–18.

Blanco, J. C. "The extinction of the wolf in Spain. Account of a scientific fraud." *Biologica* 26 (1998): 56–59.

Blanco, J. C., L. Cuesta, and S. Reig, eds. *El Lobo* (Canis lupus) *en Espana: Situacion, Problematica y Apuntes Sobre su Ecologia (The Wolf* [Canis lupus] *in Spain: Situation, Problems and Annotations on Its Ecology).* Madrid: Ministero de Agricultura Pesca y Almentacion, Icona, Publicaciones del Instituto Nacional para la Conservacion de la Naturaleza, 1990.

Blumm, M. C., M. A. Schoessler, and R. C. Beckworth. "Beyond the parity promise: struggling to save Columbia Basin salmon in the mid-1990s." *Environmental Law* 27 (1997): 21.

Bosselman, F. P. "Limitations inherent in the title to wetlands at common law." *Stanford Environmental Law Journal* 15 (1996): 247.

Botkin, D. *Discordant Harmonies.* New York: Oxford University Press, 1990.

Boyce, M. S. and E. Anderson. "Evaluating the role of carnivores in the Greater Yellowstone Ecosystem." In T. Clark and P. Curlee, eds. *Carnivores in Ecosystems.* New Haven, CT: Yale University Press, 1998.

Boyd, D. K., P. C. Paquet, S. Donelon, R. R. Ream, and C. C. White. "Transboundary movements of a recolonizing wolf population in the Rocky Mountains." In L. N. Carbyn, S. H. Fritts, and D. R. Seip, eds. *Ecology and Conservation of Wolves in a Changing World.* Edmonton, Alberta: Canadian Circumpolar Institute Occasional Publication 35 (1995): 135–140.

Braybrooke, D. *Meeting Needs.* Princeton, NJ: Princeton University Press, 1987.

Brundige, G. C. *Predation Ecology of the Eastern Coyote,* Canis latrans var., *in the Central Adirondacks.* Ph.D. dissertation. Syracuse: State University of New York, College of Environmental Science and Forestry, 1993.

Bryant, W. C. *The American Landscape: No. 1.* New York: E. Bliss, 1830.

Bubenik, A. B. "North American moose management in light of European experiences." *Proceedings of the North American Moose Conference Workshop* 8 (1972): 279–295.

Budiansky, S. *Nature's Keepers: The New Science of Nature Management.* New York: The Free Press, 1995.

Burch, W. R. "Human ecology and environmental management." In J. K. Agee and J. Darryll, eds. *Ecosystem Management for Parks and Wilderness.* Seattle: University of Washington Press, 1988, pp. 145–159.

Burch, W. R., Jr., J. Aley, B. Conover, and D. Field. *Adaptive Strategies for Natural Resource Organizations in the Twenty-First Century.* Washington, DC: Taylor Francis, 1997.

Burke, E. *A Philosophical Inquiry into the Origins of Our Ideas of the Sublime and the Beautiful.* London: R. & J. Dodsley, 1761.

Burke, E. *Reflections on the Revolution in France.* London: Dent, 1790.

Cahalane, V. H. *A Preliminary Study of Distribution and Numbers of Cougar, Grizzly and Wolf in North America.* Bronx: New York Zoological Society, 1964.

Caldwell, L. K. and K. Shrader-Frechette. *Policy for Land, Law and Ethics.* Savage, MD: Rowman & Littlefield, 1993.

California Wilderness Coalition. "California's Vanishing Forests: Two Decades of Destruction." Davis, CA: California Wildnerness Coalition, 1998.

Carson, R. *Silent Spring.* Boston: Houghton Mifflin, 1962.

Casey, D. and T. W. Clark. *Tales of the Wolf: Fifty-One Stories of Wolf Encounters in the Wild.* Moose, WY: Homestead, 1996.

Caughley, G. *Analysis of Vertebrate Populations.* London: Wiley, 1977.

Caughley, G. and A. R. E. Sinclair. *Wildlife Ecology and Management.* Oxford, UK: Blackwell Scientific, 1994.

Chambers, R. E. "Diets of Adirondack coyotes and red foxes: significance and implications." *Transactions of the 44th Annual Northeast Fish and Wildlife Conference.* Boston, 1987 (abstract only).

Charlesworth, D. and B. Charlesworth. "Inbreeding depression and its evolutionary consequences." *Annual Review of Ecology and Systematics* 18 (1987): 237–268.

Chase, A. *Playing God in Yellowstone.* New York: Harcourt Brace Jovanovich, 1987.

Chaucer, G. *The Legend of Good Women,* Prologue (1377). Oxford, UK: Clarendon, 1889.

Christensen, N. L., J. K. Agee, P. F. Brussard, J. Hughes, D. H. Knight, G. W. Minshall, J. M. Peek, S. J. Pyne, F. J. Swanson, J. W. Thomas, S. Wells, S. E. Williams, and H. A. Wright. "Interpreting the Yellowstone fires of 1988." *BioScience* 39 (1989): 678–685.

Christensen, N. L., A. M. Bartuska, J. H. Brown, S. Carpenter, C. D'Antonio, R. Francis, J. F. Franklin, J. A. MacMahon, R. F. Noss, D. J. Parsons, C. H. Peterson, M. G. Turner, and R. G. Woodmansee. "The report of the Ecological Society of America Committee on the Scientific Basis for Ecosystem Management." *Ecological Applications* 6 (1996): 665–691.

Christy v. Hodel, 857 F.2d 1324 (9 Cir. 1988).

Ciriacy-Wantrup, S. V. *Resource Conservation: Economics and Politics.* Berkeley: University of California Division of Agricultural Sciences, 1968.

Clark, C. H., and the Temporary Study Commission on the Future of the Adirondacks. *Technical Report 2: Wildlife.* 1970. Reprinted by the Adirondack Museum. Utica: Widtman Press, 1971.

Clark, S. *Animals and Their Moral Standing.* London: Routledge, 1997.

Clark, T. W. "Learning as a strategy for improving endangered species conservation." *Endangered Species Update* 13 (1996a): 5–6, 23–24.

Clark, T. W. "Appraising threatened species recovery efforts: practical recommendations." In S. Stephens and S. Maxwell, eds. *Back from the Brink: Refining the Endangered Species Recovery Process.* Sydney: Surrey Beatty & Sons, 1996b.

Clark, T. W. *Averting Extinction: Reconstructing Endangered Species Recovery.* New Haven, CT: Yale University Press, 1997a.

Clark, T. W. "Conservation biologists in the policy process: learning how to be practical and effective." In G. K. Meffe and C. R. Carroll, eds. *Principles of Conservation Biology,* 2nd ed. Sunderland, MA: Sinauer Associates, 1997b.

Clark, T. W. "Interdisciplinary problem solving in endangered species conservation: the Yellowstone grizzly bear case." In R. P. Reading and B. J. Miller, eds. *Endangered Animals: A Reference Guide to Conflicting Issues.* Westport, CT: Greenwood, 2000.

Clark, T. W. "Interdisciplinary problem solving: next steps in the Greater Yellowstone Ecosystem." *Policy Sciences* 32 (1999): 393–414.

Clark, T. W. and R. D. Brunner. "Making partnerships work in endangered species conservation: an introduction to decision process." *Endangered Species Update* 13 (1996): 1–5.

Clark, T. W., A. P. Curlee, S. C. Minta, and P. Karieva, eds. *Carnivores in Ecosystems: The Yellowstone Experience.* New Haven, CT: Yale University Press, 1999.

Clark, T. W., A. P. Curlee, and R. P. Reading. "Crafting effective solutions to the large carnivore conservation problem." Special issue. *Conservation Biology* 10 (1996a): 940–948.

Clark, T. W. and S. C. Minta. *Greater Yellowstone's Future: Prospects for Ecosystem Science, Management, and Policy.* Moose, WY: Homestead, 1994.

Clark, T. W., P. Paquet, and A. P. Curlee, eds. "Large carnivore conservation in the Rocky Mountains of the United States and Canada." Special issue. *Conservation Biology* 10 (1996b): 936–1058.

Clark, T. W., R. P. Reading, and A. L. Clarke, eds. *Endangered Species Recovery: Finding the Lessons, Improving the Process.* Washington, DC: Island Press, 1994.

Clark, T. W. and R. L. Wallace. "Understanding the human factor in endangered species recovery: an introduction to human social process." *Endangered Species Update* 15 (1998): 2–9.

Clean Water Act, 33 U.S.C. 404, sec. 1344. 1972.

Cohen, J. E. *How Many People Can the Earth Support?* New York: W. W. Norton, 1995.

Cook, K. "Night of the wolf." *Reader's Digest* 151 (1997): 114–119.

Coolidge, C. Speech, November 27, 1920. Quoted in R. Andrews, ed. *The Columbia Dictionary of Quotations.* New York: Columbia University Press, 1993, p. 736.

Costello, C. *Black Bear Habitat Ecology in the Central Adirondacks as Related to Food Abundance and Forest Management.* Master's thesis. Syracuse: State University of New York, College of Environmental Science and Forestry, 1992.

Crisler, L. 1958. *Arctic Wild.* New York: Harper, 1958.

Cromley, C. M. *Preliminary Assessment of Attitudes and Knowledge of Jackson Hole Residents Toward Grizzly Bears and Wolves in Teton County.* Jackson, WY: Northern Rockies Conservation Cooperative, 1997.

Cronon, W. *Changes in the Land: Indians, Colonists, and the Ecology of New England.* New York: Hill & Wang, 1983.

Cronon, W. "The trouble with wilderness; or, getting back to the wrong nature." In W. Cronon, ed. *Uncommon Ground: Toward Reinventing Nature.* New York: W. W. Norton, 1995a, pp. 69–90.

Cronon, W. "Introduction: In search of nature." In W. Cronon, ed. *Uncommon Ground: Toward Reinventing Nature.* New York: W. W. Norton, 1995b, pp. 23–56.

Dailey, G., ed. *Nature's Services.* Covelo, CA: Island Press, 1997.

Defenders of Wildlife. "Interior proposes reduced protection for some U.S. wolves." *Defenders* 73(4, 1998): 23.

DelGuidice, G. D., R. O. Peterson, and W. M. Samuel. "Trends of winter nutritional restriction, ticks, and numbers moose on Isle Royale." *Journal of Wildlife Management* 61 (1997): 895–903.

Demers, C. L. "Huntington Wildlife Forest wildlife observation survey." Adirondack Ecological Center, ALTEMP Report No. 13, 1998.

Deveney, P. "Title, jus publicum, and the public trust: an historical analysis." *Sea Grant Journal* 1 (1976): 13.

Dizard, J. E. *Going Wild: Hunting, Animal Rights, and the Contested Meaning of Nature,* rev. ed. Amherst: University of Massachusetts Press, 1999.

Donahue, C., Jr. "What causes fundamental legal ideas? Marital property in England and France in the thirteenth century." *Michigan Law Review* 78 (1979): 59.

Duda, M. *Public Opinion on and Attitudes Toward the Reintroduction of the Eastern Timber Wolf to Adirondack Park.* Harrisonburg, VA: Responsive Management, 1995.

Dunlap, T. R. *Saving America's Wildlife: Ecology and the American Mind, 1850–1990.* Princeton, NJ: Princeton University Press, 1988.

Easterbrook, G. *A Moment on Earth: The Coming Age of Environmental Optimism.* New York: Viking, 1995.

Eldredge, N. *Dominion.* New York: Henry Holt, 1995.

Eldredge, N. *Life in the Balance: Humanity and the Biodiversity Crisis.* Princeton, NJ: Princeton University Press, 1998.

Eliade, M. *Cosmos and History: The Myth of the Eternal Return.* New York: Garland, 1985.

Ellis, V. "Lake must rise before streams may be diverted." *Los Angeles Times,* April 19, 1991, B1.

Empire Water & Power Co. v. Cascade Town Co., 205 Fed. 123 (8 Cir. 1913).

Enck, J. W. and T. L. Brown. "Preliminary assessment of social feasibility for reintroducing gray wolves to the Adirondack Park in Northern New York." Series 00-3. Ithaca, NY: Human Dimensions Research Unit, Cornell University, 2000.

Endangered Species Act of 1973 (and amendments), 16 USC 1531 et seq.

Engelhart, S. and K. Hazard. "Wolves in the Adirondacks." *The Conservationist* (1975): 9–11.

Evers, A. *The Catskills from Wilderness to Woodstock.* Woodstock, NY: Overlook Press, 1982.

Farmer, M. C. and A. Randall. "Policies for sustainability." *Land Economics* 73 (1997): 608.

Feldman, F. and S. E. Weil. *Art Law.* Boston: Little, Brown, 1986.

Ferguson, G. *The Yellowstone Wolves: The First Year.* Helena, MT: Falcon, 1996.

Ferry v. Spokane, P. & S. Ry., 258 U.S. 314 (1922).

Field, M. E. "The evolution of the wildlife taking concept from its beginning to its culmination in the Endangered Species Act." *Houston Law Review* 21 (1984): 457.

Fischer, H., B. Snape, and W. Hudson. "Building economic incentive into the Endangered Species Act." *Endangered Species Technical Bulletin* 19(2, 1994): 4–5.

Forbes, G. J., and J. B. Theberge. "Cross-boundary management of Algonquin Park wolves." *Conservation Biology* 10 (1996): 1091–1097.

Forman, R. T. *Land Mosaics: The Ecology of Landscapes and Regions.* New York: Cambridge University Press, 1995.

Foster, D. and G. Motzkin. "Ecology and conservation in the cultural landscape of New England: lessons from nature's history." *Northeastern Naturalist* 5 (1998): 111–126.

Fox, L. B. *Ecology and Population Biology of the Bobcat,* Felis rufus *in New York.* Ph.D. dissertation. Syracuse: State University of New York, College of Environmental Science and Forestry, 1990.

Freeman, A. M., III. *The Measurement of Environmental and Resource Values: Theory and Methods.* Washington, DC: Resources for the Future, 1993.

Freud, S. *Civilization and Its Discontents.* Trans. Jean Riviere. New York: Cape & Smith, 1930.

Friend, T. "Please don't oil the animatronic warthog." *Outside* 23(5, May 1998): 100–102.

Fritts, S. H. *Wolf Depredation on Livestock in Minnesota.* Washington, DC: U.S. Department of the Interior, 1982.

Fritts, S. H. "Record dispersal by a wolf from Minnesota." *Journal of Mammalogy* 64 (1983): 166–167.

Fritts, S. H., E. E. Bangs, J. A. Fontaine, W. G. Brewster, and J. F. Gore. "Restoring wolves to the northern Rocky Mountains of the U.S." In L. N. Carbyn, S. H. Fritts, and D. R. Seip, eds. *Ecology and Conservation of Wolves in a Changing World.* Edmonton, Alberta: Canadian Circumpolar Institute Occasional Publication 35 (1995): 107–125.

Fritts, S. H., E. E. Bangs, J. A. Fontaine, M. R. Johnson, M. K. Phillips, E. D. Koch, and J. R. Gunson. "Planning and implementing a reintroduction of wolves to Yellowstone National Park and central Idaho." *Restoration Ecology* 5 (1997): 7–27.

Fritts, S. H. and L. D. Mech. "Dynamics, movements, and feeding ecology of a newly protected wolf population in northwestern Minnesota." *Wildlife Monographs* 80 (1981).

Fritts, S. H. and W. J. Paul. "Interactions of wolves and dogs in Minnesota." *Wildlife Society Bulletin* 17 (1989): 121–123.

Fritts, S. H., W. J. Paul, L. D. Mech, and D. P. Scott. "Trends and management of wolf–livestock conflicts in Minnesota." Washington, DC: U.S. Fish and Wildlife Service Resource Publication Series 181, 1992.

Fuller, T. K. "Population dynamics of wolves in north-central Minnesota." *Wildlife Monograph* 105 (1989): 1–41.

Fuller, T. K., W. E. Berg, G. L. Radde, M. S. Lenarz, and G. B. Joselyn. "A history and current estimate of wolf distribution and numbers in Minnesota." *Wildlife Society Bulletin* 20 (1992): 42–55.

Futuyma, D. J. *Evolutionary Biology,* 2nd ed. Sunderland, MA: Sinauer Associates, 1986.

Gasaway, W. C., R. O. Stephenson, J. L. Davis, P. E. K. Shepherd, and O. E. Burris. "Interrelationships of wolves, prey and man in interior Alaska." *Wildlife Monographs* 84 (1983): 1–50.

Gates, P. *History of Public Land Law Development.* Washington, DC: U.S. Government Printing Office, 1968.

Gaudreau, D. C. "The distribution of late quaternary forest regions in the Northeast: pollen data, physiography, and the prehistoric record." In G. P. Nicholas, ed. *Holocene Human Ecology in Northeastern North America.* New York: Plenum, 1988, pp. 215–256.

Geist, V. *Deer of the World.* Mechanicsburg, PA: Stackpole, 1997.

Gerrard, M. B., D. A. Ruzow, and P. Weinberg. *Environmental Impact Review in New York.* New York: Matthew Bender, 1995.

Gese, E. M. and L. D. Mech. "Dispersal of wolves *(Canis lupus)* in northeastern Minnesota, 1969–1989." *Canadian Journal of Zoology* 69 (1991): 2946–2955.

Ginsburg, J. R. and D. W. Macdonald. *Foxes, Wolves, Jackals, and Dogs.* Gland, Switzerland: International Union for the Conservation of Nature and Natural Resources, 1990.

Glover, J. M. *A Wilderness Original: The Life of Bob Marshall.* Seattle: The Mountaineers, 1986.

Gohdes, C., ed. *Hunting in the Old South: Original Narratives of the Hunters.* Baton Rouge: Louisiana State University Press, 1967.

Gould, S. J. *Ever Since Darwin.* New York: W.W. Norton, 1977.

Grumbine, R. E. "What is ecosystem management?" *Conservation Biology* 8 (1994): 27–38.

Gunderson, L. H., C. S. Holling, and S. S. Light, eds. *Barriers & Bridges: To the Renewal of Ecosystems and Institutions.* New York: Columbia University Press, 1995.

Gustafson, K. A. *Winter Metabolism and Bioenergetics of the Bobcat in New York.* M.S. thesis. Syracuse: State University of New York, College of Environmental Science and Forestry, 1984.

Gustafson, K. A. and R. H. Brocke. "Restoration of the lynx in New York: preliminary findings." *New York State Natural History Conference,* 1990 (abstract only).

Gustafson, K. A. and R. H. Brocke. "Lynx translocation in New York: ecology, biopolitics and human impacts." *Proceedings of the 5th Annual Conference of the Wildlife Society,* Buffalo, New York, 1998 (abstract only).

Hacket, D. Quote by Hank Fischer. *Casper Star Tribune,* July 7, 1993, A1.

Hairston, N. G., F. E. Smith, and L. B. Slobodkin. "Community structure, population control, and competition." *American Naturalist* 94 (1960): 421–425.

Harrison, D. J. and T. G. Chapin. "An assessment of potential habitat for eastern timber wolves in the northeastern United States and connectivity with habitat in southeastern Canada." New York: Wildlife Conservation Society, Working Paper No. 7, 1997.

Harrison, P. T. C., P. Holmes, and C. D. N. Humfrey. "Reproductive health in humans and wildlife: are adverse trends associated with environmental chemical exposure?" *Science of the Total Environment* 205 (1997): 97–106.

Hart, J. F. "Colonial land use law and its significance for the modern takings doctrine." *Harvard Law Review* 109 (1996): 1252.

Haskell, B. D., B. G. Norton, and R. Costanza. "What is ecosystem health and why should we worry about it?" In R. Costanza, B. G. Norton, and B. D. Haskell, eds. *Ecosystem Health: New Goals for Environmental Management.* Washington, DC: Island Press, 1992, pp. 3–19.

Hendrickson, J., W. L. Robinson, and L. D. Mech. "Status of the wolf in Michigan. 1973." *American Midland Naturalist* 94 (1995): 226–232.

Henshaw, R. E. "Can the wolf be returned to New York?" In F. H. Harrington and P. C. Paquet, eds. *Wolves of the World: Perspectives on Behavior, Ecology and Conservation.* Park Ridge, NJ: Noyes, 1982.

Herrick, C. L. "The mammals of Minnesota." *Geological and Natural History Survey Minnesota Bulletin* 7 (1892): 1–300.

Hicks, A., R. Inslerman, and D. Faulknham. "Wolf restoration in New York: the state's perspective." In N. Fascione and M. Cecil, compilers. *Proceedings of Defenders of Wildlife's Wolves of America Conference,* Albany, New York, November 14–16, 1996. Washington, DC: Defenders of Wildlife, 1996.

Hodgson A. "Wolf restoration in the Adirondacks? The questions of local residents." New York: Wildlife Conservation Society, Working Paper No. 8, 1997.

Holling, C. S. *Adaptive Environmental Assessment and Management.* New York: John Wiley and Sons, 1978.

Hosack, D. A. "Biological potential for eastern timber wolf re-establishment in Adirondack Park." In N. Fascione and M. Cecil, compilers. *Proceedings of Defenders of Wildlife's Wolves of America Conference,* Albany, New York, November 14–16, 1996. Washington, DC: Defenders of Wildlife, 1996.

Houck, O. A. "Why do we protect endangered species and what does that say about whether restrictions on private property to protect them constitute 'takings'?" *Iowa Law Review* 80 (1976): 297.

Houston, D. B. *The Northern Yellowstone Elk: Ecology and Management.* New York: Macmillan, 1982.

Howarth, R. B. "Sustainability under uncertainty: a deontological approach." *Land Economics* 71 (1995): 417–427.

Huffman, M. Quote by Ernie Wampler. *Jackson Hole News,* September 18, 1993, A1.

Hughes, D. *North American Indian Ecology,* 2nd ed. El Paso: Texas Western Press, 1996.

Hume, D. *An Enquiry Concerning the Principles of Morals.* L. A. Selby-Bigge and P. H. Nidditch, eds. Oxford, UK: Oxford University Press, 1986.

Huston, M. "Biological diversity, soils, and economics." *Science* 262 (1993): 1676–1680.

Iltis, H. "To the taxonomist and the ecologist, whose fight is the preservation of nature." *BioScience* 17 (1967): 886–887.

Inslerman, R. A. "Wolf restoration in New York: the state's perspective." *Pace Environmental Law Review* 15 (1998): 489–496.

International Wolf Center. *Directory of Wolf Organizations.* Ely, MN: International Wolf Center, 1982.

James, W. *Memories and Studies.* New York: Longmans Green, 1911.

Jhala, Y. V. and D. K. Sharma. "Child-lifting by wolves in eastern Uttar Pradesh, India." *Journal of Wildlife Research* 2 (1997): 94–101.

Joslin, P. "Reintroduction: the future for wolves in Alaska?" In N. Fascione and H. Ridgley, compilers. *Proceedings.* Defenders of Wildlife's Restoring the Wolf Conference (1998): 90.

Just v. Marinette County, 201 N.W.2d 761, 768 (Wis. 1972).

Kains-Jackson, C. *Our Ancient Monuments and the Land Around Them,* with a preface by Sir John Lubbock. London: E. Stock, 1880.

Kellert, S. R. "The public and the timber wolf in Minnesota." In R. E. McCabe, ed. *Transactions.* Reno, NV: North American Wildlife and Natural Resources Conference 56 (1986): 193–200.

Kellert, S. "Public views of wolf restoration in Michigan." *Transactions of the 56th North American Wildlife & Natural Resources Conference.* Reno, NV, 1991.

Kellert, S. "Concepts of nature east and west." In M. E. Soule and G. Lease, eds. *Reinventing Nature: Responses to Postmodern Deconstruction.* Washington, DC: Island Press, 1994.

Kellert, S. R., M. Black, C. R. Rush, and A. J. Bath. "Human culture and large carnivore conservation in North America." *Conservation Biology* 10 (1996): 977–990.

Kellert, S. R. and HBRS, Inc. *Public Attitudes and Beliefs About the Wolf and Its Restoration in Michigan.* Madison, WI: HBRS, Inc., 1990.

Kempton, W., J. S. Boster, and J. A. Hartley. *Environmental Values in American Culture.* Cambridge: Massachusetts Institute of Technology Press, 1995.

Kennedy, P. *Preparing for the Twenty-First Century.* New York: Random House, 1993.

Kentucky Constitution of 1850, art. XIII.

Kersley, R. H. *Goodeve's Modern Law of Personal Property,* 9th ed. London: Sweet & Maxwell, 1949.

Klein, R. A. *Wolf Recovery in the Northern Rockies: Clarifying and Securing the Common Interest.* Master's thesis. Boulder: University of Colorado, Department of Political Science, 1998.

Kluger, J. "The big (not so bad) wolves of Yellowstone back from near extinction." *Time,* January 17, 1998, 22–25.

Kolasa, J. and S. T. A. Pickett, eds. *Ecological Heterogeneity.* New York: Springer-Verlag, 1991.

Kuznik, F. "Eating themselves out of house and home." *National Wildlife* (October–November 1998): 38–43.

Lack, D. *The Natural Regulation of Animal Numbers.* Oxford, UK: Oxford University Press, 1954.

Larrère, C. "Ethics, politics, science, and the environment: concerning the natural contract." In J. B. Callicott and F. Rocha, eds. *Earth Summit Ethics: Toward a Reconstructive Postmodern Philosophy of Environmental Education.* Albany: State University of New York Press, 1996.

Lee, K. N. *Compass and Gyroscope: Integrating Science and Politics for the Environment.* Washington DC: Island Press, 1993.

Lehmann, S. *Privatizing Public Lands.* New York: Oxford University Press, 1995.

Lemonick, M. D., D. Bjerklie, A. Dorfman, and P. Dawson. "Who owns these bones?" *Time,* October 6, 1997, 74.

Leopold, A. "A biotic view of land." *Journal of Forestry* 37 (1939): 727–730.

Leopold, A. *A Sand County Almanac.* New York: Oxford University Press, 1949.

Levin, S. A. "Mechanisms for the generation and maintenance of diversity." In R. W. Hiorns and D. Cooke, eds. *The Mathematical Theory of the Dynamics of Biological Populations.* London: Academic Press, 1981, pp. 173–194.

Lewis, C. S. *The Lion, the Witch and the Wardrobe.* New York: Macmillan, 1950.

Likens, G. E., ed. *Long-term studies in ecology: approaches and alternatives.* New York: Springer-Verlag, 1989.

Litchfield, L. "Panthers in peril." *Zoo Life* 4 (1993): 42–47.

Locke, J. *Second Treatise of Government.* In P. Laslett, ed. *Locke's Two Treatises of Government.* Cambridge, UK: Cambridge University Press, 1963.

Lohr, C. and W. B. Ballard. "Historical occurrence of wolves, *Canis lupus,* in the maritime provinces." *Canadian Field Naturalist* 110 (1996): 607–610.

Lopez, B. H. *Of Wolves and Men.* New York: Scribner, 1978.

Lowenthal, D. *The Past Is a Foreign Country.* New York: Cambridge University Press, 1985.

Lucas v. South Carolina Coastal Council, 505 U.S. 1003 (1992).

Lueck, D. "The economic nature of wildlife law." *Journal of Legal Studies* 18 (1989): 291.

Lund, T. A. *American Wildlife Law.* Berkeley: University of California Press, 1980.

Lynch, K. *What Time Is This Place?* Cambridge: Massachusetts Institute of Technology Press, 1972.

Mader, T. Letter to the editor. *Casper Star Tribune,* July 28, 1993, B1.

Madonna, K. J. "The wolf in North America: defining international ecosystems vs. defining international boundaries." *Journal of Land Use & Environmental Law* 10 (1995): 305–342.

Marsh, G. P. *Man and Nature; Or, Physical Geography as Modified by Human Action.* Cambridge, MA: Harvard University Press, 1965.

Martin, C. *Keepers of the Game: Indian–Animal Relationships in the Fur Trade.* Berkeley: University of California Press, 1984.

Mattfeld, G. F. "Northeastern hardwood and spruce/fir forests." In L. K. Halls and C. House, eds. *White-Tailed Deer Ecology and Management.* Harrisburg, PA: Stackpole, 1984, pp. 305–330.

Mayr, E. "Notes on nomenclature and classification." *Systematic Zoology* 3 (1954): 86–89.

Mayr, E. *One Long Argument.* Cambridge, MA: Harvard University Press, 1991.

McDonnell, M. J. and S. T. A. Pickett, eds. *Humans as Components of Ecosystems: The Ecology of Subtle Human Effects and Populated Areas.* New York: Springer-Verlag, 1993.

McDougal, M. S., H. D. Lasswell, and R. Reisman. "The world constitutive process of authoritative decision." In M. S. McDougal and W. M. Reisman, eds. *International Law Essays: A Supplement to International Law in Contemporary Perspective.* New York: Foundation Press, 1981.

McKibben, B. *The End of Nature.* New York: Anchor, 1990.

McKibben, B. *Hope, Human and Wild.* Boston: Little, Brown, 1995.

McLaren, B. E. and R. O. Peterson. "Wolves, moose and tree rings on Isle Royale." *Science* 266 (1994): 1555–1558.

McNamee, T. *The Return of the Wolf to Yellowstone.* New York: Henry Holt, 1997.

McNulty, S. A. *Cover Type, Logging Disturbance, and Recruitment of White-Tailed Deer in the Adirondacks.* Master's thesis. Syracuse: State University of New York, College of Environmental Science and Forestry, 1997.

Mech, L. D. *The Wolves of Isle Royale.* U.S. National Park Fauna Series 7. Washington, DC: U.S. Government Printing Office, 1966.

Mech, L. D. *The Wolf: The Ecology and Behavior of an Endangered Species.* New York: Doubleday, 1970.

Mech, L. D. "Age, season, distance, direction, and social aspects of wolf dispersal from a Minnesota pack." In B. D. Chepko-Sade and Z. Halpin, eds. *Mammalian Dispersal Patterns* (1987): 55–74.

Mech, L. D. "The challenge and opportunity of recovering wolf populations." *Conservation Biology* 9 (1995a): 270–278.

Mech, L. D. Letter to the editor. *Casper Star Tribune,* January 19, 1995b, E6.

Mech, L. D. "What do we know about wolves and what more do we need to learn?" In L. N. Carbyn, S. H. Fritts, and D. R. Seip, eds. *Ecology and Conservation of Wolves in a Changing World*. Edmonton, Alberta: Canadian Circumpolar Institute Occasional Publication 35 (1995c): 537–545.

Mech, L. D. "A new era for carnivore conservation." *Wildlife Society Bulletin* 24 (1996): 397–401.

Mech, L. D. "Estimated costs of maintaining a recovered wolf population in agricultural regions of Minnesota." *Wildlife Society Bulletin* 26 (1998): 817–822.

Mech, L. D., L. G. Adams, T. J. Meier, J. W. Burch, and B. W. Dale. *The Wolves of Denali*. Minneapolis: University of Minnesota Press, 1998.

Mech, L. D. and L. D. Frenzel Jr. *Ecological Studies of the Timber Wolf in Northeastern Minnesota*. St. Paul, MN: North Central Forest Experiment Station NC-52, 1971.

Merchant, C. *Ecological Revolutions: Nature, Gender, and Science in New England*. Chapel Hill: University of North Carolina Press, 1989.

Messier, F. "Solitary living and extra-territorial movements of wolves in relation to social status and prey abundance." *Canadian Journal of Zoology* 63 (1985): 239–245.

Meyers, M. Letter to the editor. *Sunday Republican* (Springfield, MA), December 7, 1997, B3.

Michigan Department of Natural Resources. *Michigan Gray Wolf Recovery and Management Plan*. Lansing: Michigan Department of Natural Resources, 1997.

Midgley, M. *Beast and Man*. Ithaca, NY: Cornell University Press, 1978.

Midgley, M. *Animals and Why They Matter*. Athens: University of Georgia Press, 1984.

Mighetto, L. *Wild Animals and American Environmental Ethics*. Phoenix: University of Arizona Press, 1991.

Miller, A. *Environmental Problem Solving: Psychosocial Barriers to Adaptive Change*. New York: Springer, 1999.

Minnesota Poll. *Wolves*. St. Paul: Minnesota Public Radio, KARE 11, and *Pioneer Press*, 1998.

Mitchell, J. H. *Ceremonial Time: Fifteen Thousand Years on One Square Mile*. Reading, MA: Addison-Wesley, 1997.

Mladenoff, D. J. and T. A. Sickley. "Assessing potential grey wolf restoration in the northeastern United States: a spatial prediction of favorable habitat and potential population levels." *Journal of Wildlife Management* 62 (1998): 1–10.

Morris, R. B. *Studies in the History of American Law*. New York: Columbia University Press, 1930.

Mowat, F. *Never Cry Wolf*. New York: Dell, 1963.

Mulholland, M. T. "Territoriality and horticulture: a perspective for prehistoric southern New England." In G. P. Nicholas, ed. *Holocene Human Ecology in Northeastern North America*. New York: Plenum, 1988, pp. 137–166.

National Environmental Policy Act of 1969 (NEPA), 42 USC 4321 et seq.

Nelson, M. E. and L. D. Mech. "Deer social organization and wolf predation in northeastern Minnesota." *Wildlife Monographs* 77 (1981): 1–53.

Nelson, R. *Heart and Blood: Living with Deer in America*. New York: Knopf, 1997.

Nelson, W. *Manwood's Treatise on the Law of the Forest*, 4th ed. London: C. Nutt, 1717.

Nesslage, G. "Population dynamics of white-tailed deer at the Huntington Wildlife Forest." Adirondack Ecological Center Special Report No. 143. Syracuse, NY, 1999.

New York State Constitution. *McKinney's Consolidated Laws of New York Annotated*.

Book 2 Constitution. Article XIV: Conservation. St. Paul, MN: West Publishing Company, 1987, pp. 555–569.

New York State Environmental Conservation Law. *McKinney's Consolidated Laws of New York Annotated.* Book 17-1/2 Environmental Conservation Law ECL. Article 11, Fish and Wildlife. §11-0101 to §11-2307. St. Paul, MN: West Publishing Company, 1997, pp. 332–618.

Ng, Y.-K. "What should we do about future generations? The impossibility of Parfit's Theory X." *Economics and Philosophy* 5 (1989): 235–253.

Norgaard, R. "Economic indicators of resource scarcity: a critical essay." *Journal of Environmental Economics and Management* 19 (1990): 19–25.

Norton, B. G. *Why Preserve Natural Variety?* Princeton, NJ: Princeton University Press, 1987.

Norton, B. G. "Commodity, amenity and morality." In E. O. Wilson, ed. *Biodiversity.* Washington, DC: National Academy Press, 1988.

Norton, B. G. "Thoreau's insect analogies: or, why environmentalists hate mainstream economists." *Environmental Ethics* 13 (1991a): 234–251.

Norton, B. *Toward Unity Among Environmentalists.* New York: Oxford University Press, 1991b.

Norton, B. G. "Economists' preferences and the preferences of economists." *Environmental Values* 3 (1994): 311–332.

Norton, B. G. "Ecology and opportunity: intergenerational equity and sustainable options." In A. Dobson, ed. *Fairness and Futurity.* Oxford, UK: Oxford University Press, 1999.

Norton, B. G., R. Costanza, and R. C. Bishop. "The evolution of preferences: why 'sovereign' preferences may not lead to sustainable policies and what to do about it." *Ecological Economics* 24 (1998): 193–211.

Norton, B. G. and M. A. Toman. "Sustainability: ecological and economic perspectives." *Land Economics* 73 (1997): 553–568.

Norton, B. G. and R. E. Ulanowicz. "Scale and biodiversity policy: a hierarchical approach." *Ambio* 21 (1992): 244–249.

Nowak, R. M. *North American Quaternary Canis.* Monograph No. 6. Lawrence: Museum of Natural History, University of Kansas, 1979.

Nowak, R. M. "Another look at wolf taxonomy." In L. N. Carbyn, S. H. Fritts, and D. R. Seip, eds. *Ecology and Conservation of Wolves in a Changing World.* Edmonton, Alberta: Canadian Circumpolar Institute Occasional Publication 35 (1995): 375–397.

Nowak, R. M., M. K. Phillips, G. V. Henry, W. C. Hunter, and R. Smith. "The origin and fate of the red wolf." In L. N. Carbyn, S. H. Fritts, and D. R. Seip, eds. *Ecology and Conservation of Wolves in a Changing World.* Edmonton, Alberta: Canadian Circumpolar Institute Occasional Publication 35 (1995): 409–415.

O'Driscoll, P. "Tragedy hunts wolves released in Southwest." *USA Today,* November 9, 1998, A3.

Overholt, T. W. and J. B. Callicott. *Clothed-in-fur and Other Tales: An Introduction to the Ojibwa World View.* Lanham: MD: University of Press of America, 1982.

Palila v. Hawaii Dept. of Land and Nat. Res., 471 F. Supp. 985 (D. Ha. 1979), aff'd. 639 F.2d 495 (9 Cir. 1981).

Paquet, P., J. Strittholt, and N. L. Staus. *Wolf Reintroduction Feasibility in the Adirondack Park*. Corvallis, OR: Conservation Biology Institute, 1999.

Parker, W. T. and M. K. Phillips. "Application of the experimental population designation to the recovery of endangered red wolves." *Wildlife Society Bulletin* 19 (1991): 73–79.

Partridge, E. "Nature as a moral resource." *Environmental Ethics* 6 (1984): 101–130.

Partridge, E. "In search of sustainable values." Paper delivered at an international conference, "Reflections on Discounting," Vilm Island, University of Greifswald, Germany, May 28, 1999. Publication of the conference papers is being negotiated by sponsors, but it is available at <www.igc.org/gadfly/papers/sustain.htm>.

Passmore, J. *Man's Responsibility for Nature*. New York: Scribner, 1974.

Penn Central Trans. Co. v. City of New York, 377 N.Y.S.2d 20.36 (App. Div., 1st Dept., 1975).

Penn Central Trans. Co. v. City of New York, 438 U.S. 104 (1978).

Peterson, R. O. "Wolf ecology and prey relationships on Isle Royale." U.S. Park Service Science Monographs Series 11, 1977.

Peterson, R. O. "Wolf–moose interaction on Isle Royale: the end of natural regulation?" *Ecological Applications* 9 (1999): 10–16.

Peterson, R. O. "Wolves as interspecific competitors in canid ecology." In Carbyn, L. N., S. H. Fritts, and D. R. Seip, eds. *Ecology and Conservation of Wolves in a Changing World*. Edmonton, Alberta: Canadian Circumpolar Institute Occasional Publication 35 (1995): 315–323.

Peterson, R. O. and R. E. Page. "Detection of moose in midwinter from fixed-wing aircraft over dense forest cover." *Wildlife Society Bulletin* 21 (1993): 80–86.

Peterson, R. O., N. J. Thomas, J. M. Thurber, J. A. Vucetich, and T. A. Waite. "Population limitation and the wolves of Isle Royale." *Journal of Mammalogy* 79 (1998): 828–841.

Peterson, R. O., J. D. Woolington, and T. N. Bailey. "Wolves of the Kenai Peninsula, Alaska." *Wildlife Monograph* 88, 1984.

Pickett, S. T. A. and V. T. Parker. "Avoiding the old pitfalls: opportunities in a new discipline." *Restoration Ecology* 2 (1994): 75–79.

Pickett, S. T. A., V. T. Parker, and P. Fiedler. "The new paradigm in ecology: implications for conservation biology above the species level." In P. Fiedler and S. Jain, eds. *Conservation Biology: The Theory and Practice of Nature Conservation, Preservation, and Management*. New York: Chapman & Hall, 1992, pp. 65–88.

Pimlott, D. H. "Wolf predation and ungulate populations." *American Zoologist* 7 (1967): 267–278.

Post, E., R. O. Peterson, N. C. Stenseth, and B. E. McLaren. "Ecosystem consequences of wolf behavioural responses to climate." *Nature* 401 (1999): 905.

Powell, R. R. B. "The relationship between property rights and civil rights." *Hastings Law Journal* 15 (1963): 135, 142–143.

"Protecting the public interest in art." *Yale Law Review* 91 (1981): 121.

Putz, F. E. "Florida's forests in the year 2020 and deeper into the homogocene." *Journal of the Public Interest Environmental Conference* 1 (1997): 91–97.

Quammen, D. "Planet of weeds." *Harper's Magazine* 297 (October 1998): 57–69.

Ralls, K., J. D. Ballou, and A. Templeton. "Estimates of lethal equivalents and the cost of inbreeding in mammals." *Conservation Biology* 2 (1988): 185–193.

Randall v. Kreiger, 90 U.S. 137, 148 (1874).

Ratti, J. T., M. Weinstein, J. M. Scott, P. Avsharian, A. Gillesberg, C. A. Miller, M. M. Szepanski, and L. Bomer. Final draft: *Feasibility Study on the Reintroduction of Gray Wolves to the Olympic Peninsula.* Lacey, WA: U.S. Fish and Wildlife Service, 1999. <www.nctc.fws.gov/library/Pubs9/graywolf_olympic.doc>.

Rawls, J. *A Theory of Justice.* Boston: Harvard University Press, 1971.

Rawls, J. *Political Liberalism.* New York: Columbia University Press, 1996.

Reading, R. P., T. W. Clark, and B. Griffith. "The influence of valuation and organizational considerations on the success of rare species translocations." *Biological Conservation* 70 (1996): 217–225.

Reiger, J. F. *American Sportsmen and the Origins of Conservation.* New York: Winchester, 1975.

Robbins, J. "In 2 years, wolves reshaped Yellowstone." *New York Times,* December 30, 1997, D1.

Rogers, Will. *The Autobiography of Will Rogers.* 1949. Quoted in R. Andrews, ed. *The Columbia Dictionary of Quotations.* New York: Columbia University Press, 1993, 738.

Rolston, H., III. "Is there an ecological ethic?" *Ethics* 85 (1975): 93–109.

Romme, W. H., M. G. Turner, L. L. Wallace, and J. S. Walker. "Aspen, elk, and fire in northern Yellowstone Park." *Ecology* 76 (1995): 2097–2106.

Roosevelt, T. R. *Hunting Trips of a Ranchman and the Wilderness Hunter: An Account of the Big Game of the United States and Its Chase with Horse, Hound and Rifle.* New York: G. P. Putman, 1900.

Rozzi, R., J. J. Armesto, S. T. A. Pickett, S. Lehmann, and F. Massardo. "Recuperando el vínculo entre la ciencia y la etica: una aproximición ecológica (Recovering the tie between science and ethics: an ecological approximation)." In G. Montenegro and B. Timmermann, eds. *Aspectos Ambientales, Ideológicos, Eticos y Politicos en el Debate de Bioprospección de Recursos Genéticos en Chile. Noticiero de la Sociedad de Biología de Chile* 5 (1997): 71–75.

Russell, E. W. B. "Discovery of the subtle." In M. J. McDonnell and S. T. A. Pickett, eds. *Humans as Components of Ecosystems: The Ecology of Subtle Human Effects and Populated Areas.* New York: Springer-Verlag, 1993, pp. 81–90.

Sage, R. W., Jr. "A comprehensive model: white-tailed deer population dynamics in the Adirondack ecosystem." Unpublished data, Adirondack Ecological Center, 1998.

Sagoff, M. *The Economy of the Earth.* New York: Cambridge University Press, 1988.

San Francisco Examiner, "Salvation is at hand" (editorial). September 3, 1998, A-20.

Saunders, D. A., R. J. Hobbs, and C. R. Margules. "Biological consequences of ecosystem fragmentation: a review." *Conservation Biology* 5 (1991): 18–32.

Sax, J. L. "The Public Trust doctrine in natural resources law: effective judicial intervention." *Michigan Law Review* 68 (1970): 471.

Sax, J. L. "Liberating the Public Trust doctrine from its historical shackles." *University of California Davis Law Review* 14 (1980): 185.

Sax, J. L. "Is anyone minding Stonehenge? The origins of cultural property protection in England." *California Law Review* 78 (1990a): 1543.

Sax, J. L. "Heritage preservation as a public duty: the Abbé Grégoire and the origins of an idea." *Michigan Law Review* 88 (1990b): 1142.

Sax, J. L. "Property rights and the economy of nature: understanding *Lucas v. South Carolina Coastal Council.*" *Stanford Law Review* 45 (1993): 1433.

Sax, J. L. "Inching toward a land ethic," 1987–1988 *Annual Report.* Ann Arbor: School of Natural Resources, University of Michigan, 1998.

Sax, J. L. *Playing Darts with a Rembrandt: Public and Private Rights in Cultural Treasures.* Ann Arbor: University of Michigan Press, 1999.

Scarce, R. "What do wolves mean? Conflicting social constructions of *Canis lupus* in 'Bordertown.'" *Human Dimensions of Wildlife* 3(3, Fall 1998): 26–45.

Schadler, C. L. *Eastern Timber Wolf Recovery in Michigan: Biological and Social Perspectives.* Unpublished master's thesis. Keene, NH: Antioch New England Graduate Library, 1994.

Schmitt, P. J. *Back to Nature: The Arcadian Myth in Urban America.* Baltimore: Johns Hopkins University Press, 1969.

Schön, D. A. *The Reflective Practitioner: How Professionals Think in Action.* New York: Basic Books, 1983.

Schumacher, E. F. *Good Work.* New York: Harper & Row, 1979.

Schwartz, C. C., K. J. Hundertmark, and T. H. Spraker. "An evaluation of selective bull moose harvest on the Kenai Peninsula, Alaska." *Alces* 28 (1992): 1–13.

Sen, A. "Legal rights and moral rights." *Ratio Juris* 9(2, 1996): 153–167.

Shuey, M. *Using a Meaning-Based Approach for Exploring the Underlying Belief Systems That Drive Human–Wildlife Conflicts.* Master's thesis. Clemson, SC: Department of Parks, Recreation, and Tourism Management, Clemson University, 1997.

Simon, J. *The Ultimate Resource.* Princeton, NJ: Princeton University Press, 1981.

Slater, D. "Signs of the wild." *Sierra* (September–October 1999): 30.

Smith, C. "How news media cover disasters: the case of Yellowstone." In P. S. Cook, D. Gomery, and L. W. Lichty, eds. *The Future of News: Television–Newspapers–Wire Services–Newsmagazines.* Baltimore: Johns Hopkins University Press, 1992, pp. 223–240.

Solow, R. M. "Intergenerational equity and exhaustible resources." *Review of Economic Studies* Symposium Issue (1974): 29–45.

Solow, R. M. "Sustainability: an economist's perspective." In R. Dorfman and N. Dorfman, eds. *Economics of the Environment,* 3rd ed. New York: W.W. Norton, 1993, pp. 179–187.

Southview Associates v. Bongartz, 980 F.2d 84 (2 Cir. 1992).

State Dept. of Parks v. Idaho Dept. of Water Administration, 96 Idaho 440, 530 P.2d 924 (1974).

State ex rel. Sackman v. State Fish & Game Comm'n, 438 P.2d 663 (1968).

Stephenson, R. O., W. Ballard, C. Smith, and K. Richardson. "Wolf biology and management in Alaska, 1981–1991." In L. N. Carbyn, S. H. Fritts, and D. R. Seip, eds. *Ecology and Conservation of Wolves in a Changing World.* Edmonton, Alberta: Canadian Circumpolar Institute Occasional Publication 35 (1995): 43–54.

Steptoe Live Stock Co. v. Gulley, 53 Nev. 163, 295 Pac. 772 (1931).

Sterste, C. Letter to the editor. *Daily Hampshire Gazette,* February 23, 1998, 10.

Stevens, W. K. *Miracle Under the Oaks: The Revival of Nature in America.* New York: Pocket Books, 1995.

Sturges v. Crownshield, 17 U.S. (4 Wheat.) 122 (1819).

Sullivan, R. *The Meadowlands: Wilderness Adventures at the Edge of a City.* New York: Scribner, 1998.

Surber, T. *The Mammals of Minnesota.* St. Paul: Minnesota Department of Conservation, 1932.

Tennyson, A., Lord. "Northern Farmer, Old Style" (1861). In C. Ricks, ed. *The Poems of Tennyson,* Vol. 2. Berkeley: University of California Press, 1987, p. 619.

Theberge, J. and M. Theberge. *Wolf Country.* Toronto: McClelland & Stewart, 1998.

Thiel, R. P. *The Timber Wolf in Wisconsin: The Death and Life of a Majestic Predator.* Madison: University of Wisconsin Press, 1993.

Thiel, R. and J. Hammill. "Wolf specimen records in Upper Michigan, 1960–1986." *Jack Pine Warbler* 66 (1988).

Thoreau, H. D. *The Journal of Henry David Thoreau* (March 23, 1856), Vol. VIII. B. Torrey and F. H. Allen, eds. Layton, UT: Gibbs Smith, 1984.

Thoreau, H. D. *Walden.* Philadelphia: Running Press, 1990. (Originally published 1854).

Toullier, C. B. M. *Le droit civil francais suivant l'ordre du Code (French civil law according to the Code)* (4th ed., rev. and corr.). Paris: Waree, 1824–1837.

Town of Newcomb. *A Comprehensive Plan.* 1990.

Turner J. F. and J. C. Rylander. "Conserving endangered species on private lands." *Land and Water Law Review* 32 (1997): 571, 577.

Underwood, H. B. *Population Dynamics of a Central Adirondack Deer Herd: Responses to Intensive Population and Forest Management.* Master's thesis. Syracuse: State University of New York, College of Environmental Science and Forestry, 1986.

Underwood, H. B. *Population Dynamics of White-Tailed Deer in the Central Adirondack Mountains of New York: Influences of Winter, Harvest, and Population Abundance.* Ph.D. dissertation. Syracuse: State University of New York, College of Environmental Science and Forestry, 1990.

United States Fish and Wildlife Service. *Recovery Plan for the Eastern Timber Wolf.* Minneapolis and St. Paul, MN: United States Fish and Wildlife Service, 1978.

United States Fish and Wildlife Service. *Northern Rocky Mountain Wolf Recovery Plan.* Denver: United States Fish and Wildlife Service, 1987.

United States Fish and Wildlife Service. *Recovery Plan for the Eastern Timber Wolf,* rev. ed. Minneapolis and St. Paul, MN: United States Fish and Wildlife Service, 1992.

United States Fish and Wildlife Service. *Final Environmental Impact Statement: The Reintroduction of Gray Wolves to Yellowstone National Park and Central Idaho.* Denver: United States Fish and Wildlife Service, 1994.

USA Today. Advertisement. August 10, 1995.

USA Today. Advertisement. August 3, 1998.

Valentino, P. C. "Of wolves, cows, and humans." In N. Fascione and H. Ridgley, compilers. *Proceedings, Defenders of Wildlife's Restoring the Wolf Conference* (1998): 47–53.

Van Camp, J. and R. Gluckie. "A record long distance move by a wolf." *Journal of Mammalogy* 60 (1979): 236.

Vance, W. R. "The quest for tenure in the United States." *Yale Law Review* 33 (1924): 248, 249.

Varley, J. D. and W. G. Brewster, eds. "Wolves for Yellowstone?" A report to the United

States Congress. Volume III, *Executive Summaries*. Washington, DC: U.S. Government Printing Office, 1992.

Viscome, L. "The Adirondack Park: one of a kind." *The Conservationist* 46(6, 1992): 8–13.

Wallace R. L. and T. W. Clark. "Solving problems in endangered species conservation: an introduction to problem orientation." *Endangered Species Update* 16 (1999): 28–34.

Waller, D. M. "Getting back to the right nature: a reply to Cronon's 'The Trouble with Wilderness.'" In J. B. Callicott and M. P. Nelson, eds. *The Great New Wilderness Debate*. Athens: University of Georgia Press, 1998, pp. 540–567.

Warburton v. White, 176 U.S. 484 (1899).

Wayne, R. K. and S. M. Jenks. "Mitochondrial DNA analysis implying extensive hybridization of the endangered red wolf, *Canis rufus*." *Nature* 351 (1991): 565–568.

Webster, D. "Can Ted Turner keep the west wild?" *Audubon* (January–February 1999): 48–56.

Weise, T. F., W. L. Robinson, R. A. Hook, and L. D. Mech. "An experimental transloca-tion of the eastern timber wolf." *Audubon Conservation Report* 5 (1975).

Whittaker, R. H. "Evolution of diversity in plant communities." In G. M. Woodwell and H. H. Smith, eds. *Diversity and Sustainability in Ecological Systems*. Brookhaven National Laboratory Publication No. 22, Springfield, VA: Clearinghouse for Federal Scientific and Technical Information, 1969, pp. 178–196.

Williams, T. "Management by majority." *Audubon* 101 (1999): 40–49.

Wilson, E. O. *Biophilia*. Cambridge, MA: Harvard University Press, 1984.

Wilson, E. O. *The Diversity of Life*. Cambridge, MA: Harvard University Press, 1992.

Wilson, E. O. and W. L. Brown. "The subspecies concept and its taxonomic applica-tion." *Systematic Zoology* 2 (1953): 97–111.

Wilson, M. A. "The wolf in Yellowstone: science, symbol, or politics? Deconstructing the conflict between environmentalism and wise use." *Society & Natural Resources* 10 (1997): 453–468.

Wisconsin Draft Wolf Management Plan. Madison: Wisconsin Department of Natural Resources, 1998.

Wolf, C. "Markets, justice, and the interests of future generations." *Ethics and the Envi-ronment* 1(2, 1996): 153–175.

Wolf, C. "Human Population Growth." In Dale Jamieson, ed., *The Blackwell Compan-ion to Environmental Philosophy*. Oxford: Blackwell Publishers, 2000.

Wolok, M. S. "Experimenting with experimental populations." *Environmental Law Reporter* 26 (1996): 10018.

Wood, P. "Biodiversity as the source of biological resources." *Environmental Values* 6 (1997): 251–268.

Wydeven, A. P., T. K. Fuller, W. Weber, and K. Macdonald. "The potential for wolf recovery in the northeastern United States via dispersal from southeastern Canada." *Wildlife Society Bulletin* 26 (1998): 776–784.

Wyoming Farm Bureau v. Babbitt, 987 F Supp. 1349 (D. Wyoming, 1997).

Wyoming Farm Bureau v. Babbitt, U.S. Court of Appeals, 10th District U.S. District Court (Wyoming) DC No. 94-CV-286 (January 13, 2000).

Yellowstone National Park, U.S. Fish and Wildlife Service, University of Wyoming, University of Idaho, Interagency Grizzly Bear Study Team, and University of Minnesota Cooperative Park Studies Unit. *Wolves for Yellowstone? A Report to the U.S. Congress.* Vol. I *Executive Summaries;* Vol. II *Research and Analysis,* 1990.

Young, S. P. and E. A. Goldman. *The Wolves of North America.* Toronto, Ontario: General Publishing Company, 1944.

About the Contributors

TIMOTHY CLARK is a professor adjunct in the School of Forestry and Environmental Studies and a fellow in the Institution for Social and Policy Studies, Yale University. He is also board president of the Northern Rockies Conservation Cooperative in Jackson, Wyoming. He applies interdisciplinary methods in courses and projects in natural resource policy and management. His current policy-oriented work includes workshops on policy sciences problem solving for government and conservation groups in the United States and Australia, case analyses of koala and endangered species conservation in Australia, studies of sustainability, ecosystem management, large carnivores, national park management, and endangered species in the United States. His recent book *Averting Extinction: Reconstructing Endangered Species Recovery* is one application. He has published more than 250 papers and 10 books and monographs. He is senior editor of *Carnivores in Ecosystems* (1999).

JAN E. DIZARD is the Charles Hamilton Houston Professor of American Culture, Department of Anthropology–Sociology, Amherst College. He received his Ph.D. in sociology from the University of Chicago and has taught at the University of California, Berkeley, and, since 1969, at Amherst College. Dizard is the author of *Going Wild: Hunting, Animal Rights, and the Contested Meaning of Nature* (1994) and books and articles on the American family, race relations, and hunting.

STRACHAN DONNELLEY is director of the Humans and Nature Program at the Hastings Center in Garrison, New York. Dr. Donnelley received his B.A. from Yale University, attended Oxford University, and received his M.A. and Ph.D. from the New School for Social Research. Among research projects he has directed at the center are the Ethics of Animal Research and Animal Biotechnology. In the Humans and Nature Program he directs three current projects:

303

"Chicago Regional Planning: Nature, Polis, and Ethics," "Trout, Salmon, and Rivers: Saving the Human and Natural Future," and "Revisiting Nature: The Legacies of Charles Darwin and Aldo Leopold" (the ethics of public responsibility). A former president of the Hastings Center, he also continues to be involved in the center's international work in biomedical and environmental ethics and directs a Families and Traumatic Brain Injuries project.

NILES ELDREDGE has been an active research paleontologist on the staff of the American Museum of Natural History since 1969. He was trained under Norman D. Newell in the combined Columbia University–American Museum graduate studies program from 1965 to 1969.

Dr. Eldredge has devoted his entire career to effecting a better fit between evolutionary theory and the fossil record. His specialty as a systematist has been trilobite and other extinct arthropods. In 1972 with Stephen Jay Gould he announced the theory of punctuated equilibria. Since then he has developed his views on the hierarchical structure of living systems and the nature of the relationship between ecology and evolution. In recent years he has focused on the mass extinctions of the geological past and their implications for understanding the modern biodiversity crisis and future human ecological and evolutionary prospects.

Dr. Eldredge has published more than 160 scientific articles, books, and reviews. His recent books include *Reinventing Darwin* (1995), *Dominion* (1995), *Life in the Balance* (1998), and *The Pattern of Evolution* (1998).

NINA FASCIONE has a master's degree in conservation anthropology from the University of Maryland. Her professional work and education have focused on the biological and sociological aspects of wildlife conservation. Fascione is currently director of carnivore conservation at Defenders of Wildlife, where she manages several of the organization's recovery programs for endangered species. She served for six years as cochair of the American Zoo and Aquarium Association's Bat Advisory Group and was a member of the International Union for the Conservation of Nature and Natural Resources/Species Survival Commission Fruit Bat Specialist Group. She has authored more than two dozen journal articles, book chapters, and technical reports covering various topics in wildlife science and conservation.

ANNE-MARIE GILLESBERG is an associate of Dr. Timothy Clark at the Northern Rockies Conservation Cooperative. She earned degrees in zoology and psychology from the University of Washington in 1984. Before continuing her education at the Yale School of Forestry and Environmental Studies, Ms. Gillesberg worked with a number of academic institutions, government agencies, and nongovernment organizations over a period of 12 years on studies of wildlife–habitat relationships. Her experience is largely field based, which she considers an asset, particularly when coupled with her more recent focus on the human dimensions of conservation biology relative to wolf reintroduction to Yellowstone, the Adirondacks, and the Olympic Peninsula. Insights gained through academic study and field research have improved her approach to such politically sensitive and highly contentious conservation and management issues and her ability to promote understanding and effect change.

ROBERT A. INSLERMAN is a 31-year employee with the New York State Department of Environmental Conservation's Division of Fish, Wildlife and Marine Resources (Bureau of Wildlife) and the regional wildlife manager for the Department of Environmental Conservation's Region 5, an eight-county, 10,000-square-mile area that encompasses the eastern half of the Adirondacks from the Mohawk River on the south to the Canadian border on the north. His major responsibilities include overseeing, supervising, and administering the Region 5 wildlife program in six major program areas: habitat protection, environmental management, species management, public use, extension services, and administration. Major activities include developing and implementing various big game, fur-bearer, migratory bird, and endangered species surveys and management projects; acquiring wetlands; responding to nuisance, distressed, and diseased wildlife complaints; providing outreach and extension services including sport education programs; and budgeting, developing work plans, and prioritizing actions.

Mr. Inslerman received an associate in applied science degree in forestry from Paul Smith's College in 1965 and a B.S. in wildlife management from Cornell University in 1967.

He is a member of the New York Chapter, Northeast Section, and Parent Chapter of the Wildlife Society. He served as secretary to the New York Chapter from 1982 to 1985 and as president from 1986 to 1987. He has also served on various committees for the Wildlife Society at all levels. His interests include gardening, hunting, fishing, trapping, canoeing, bottle collecting, and making maple syrup. He lives with his wife and two daughters in Upper St. Regis, New York.

DANIEL KEMMIS, director of the Center for the Rocky Mountain West, is the former mayor of Missoula, a former speaker and minority leader of the Montana House of Representatives, and a four-term Montana legislator.

Mr. Kemmis was the first chair of the National League of Cities' Leadership Training Council; he is currently a member of the Board of Directors of the Kettering Foundation and the Pew Partnership for Civic Change. He is a fellow of the Dallas Institute and a member of the advisory boards of the Snake River Institute and of the Brookings Institution's Center on Urban and

Metropolitan Policy. In April 1998 he was appointed to the American Heritage Rivers Initiative Advisory Committee by President Clinton. He is the author of two books: *Community and the Politics of Place* (1990) and *The Good City and the Good Life* (1995).

Mr. Kemmis has had articles published in national magazines and journals on such topics as the city center, community and community building, and the economy and politics of the West. He was recognized by the *Utne Reader* in 1995 as one of its 100 Visionaries. In 1997, President Clinton awarded Mr. Kemmis the Charles Frankel Prize for outstanding contribution to the field of the humanities. Also in 1997, he received the Society for Conservation Biology's Distinguished Achievement Award for Social, Economic and Political Work. In 1998, the Center of the American West awarded him the Wallace Stegner Prize for sustained contribution to the cultural identity of the West. Mr. Kemmis is a graduate of Harvard University and the University of Montana Law School.

STEPHEN R. KENDROT received his M.S. degree in 1998 from the State University of New York College of Environmental Science and Forestry, where he studied the ecology of eastern coyotes in agricultural landscapes. He served as northeastern field representative for Defenders of Wildlife from June 1997 until May 1999. He is currently employed by the U.S. Department of Agriculture/Animal and Plant Health Inspection Service/Wildlife Services in Virginia.

L. DAVID MECH is a senior research scientist for the Biological Resources Division of the U.S. Geological Survey (formerly Division of Endangered Wildlife Research, U.S. Fish and Wildlife Service). With a B.S. in wildlife conservation from Cornell University in 1958 and a Ph.D. in wildlife ecology from Purdue University in 1962, he has studied wolves and their prey in several areas of the world for 40 years and published numerous books and articles about them. He has chaired the International Union for the Conservation of Nature and Natural Resources World Conservation Union's Wolf Specialist Group since 1978 and founded the International Wolf Center (www.wolf.org) in Ely, Minnesota, in 1985.

MARY MIDGLEY is a professional philosopher whose special interests are the relationships of humans with the rest of nature (particularly the status of animals), the sources of morality, and the relationship between science and religion (particularly in cases in which science becomes a religion). Until retirement she was a senior lecturer in philosophy at the University of Newcastle on Tyne in England, where she still lives. She is a widow. Her husband was another philosopher, Geoffrey Midgley, and she has three sons. Her best-known books are *Beast and Man* (1979), *Animals and Why They Matter* (1983), *Science as Salvation* (1992), *Wickedness* (1984), and, most recently, *Utopias, Dolphins and Computers* (1996).

BRYAN G. NORTON received his Ph.D. in philosophy from the University of Michigan in 1970. Currently professor of philosophy in the School of Public Policy, Georgia Institute of Technology, he writes on intergenerational equity, sustainability theory, biodiversity policy, and valuation methods. He is author of *Why Preserve Natural Variety?* (1987) and *Toward Unity Among Environmentalists* (1991), editor of *The Preservation of Species* (1986), and coeditor of *Ecosystem Health: New Goals for Environmental Management* (1992) and *Ethics on the Ark* (1995). He has contributed to journals in several fields, including philosophy, biology, ecology, economics, and environmental management. He has completed a term as a charter member of the Environmental Economics Advisory Committee of the Environmental Protection Agency Science Advisory Board.

ERNEST PARTRIDGE is a research associate in the Department of Philosophy, University of California, where he is currently studying disequilibrium ecology under a research grant from the National Science Foundation. His areas of interest include environmental ethics, policy analysis, moral philosophy, and applied ethics, with a particular interest in the issue of the responsibility to future generations. In recent and continuing visits to Russia, Dr. Partridge has established productive and ongoing communication and cooperation with international scholars and scientists involved in global environmental issues. He is the editor of the anthology *Responsibilities to Future Generations.* He maintains a Web site, "The Online Gadfly" (www.igc.org/gadfly), which contains his recent publications and news and opinion about environmental ethics and policy.

ROLF O. PETERSON is a professor of wildlife ecology at Michigan Technological University. He has been involved with research on wolves for more than 30 years, including studies in Alaska, Michigan, and Minnesota, and has consulted on wolf recovery in Scandinavia and the Yellowstone area. He serves on the Board of Directors of the International Wolf Center and is team leader for the Gray Wolf (Eastern Population) Recovery Team appointed by the U.S. Fish and Wildlife Service. He lives in Upper Michigan, where wolves became reestablished in the 1990s.

STEWARD T. A. PICKETT is a senior scientist at the Institute of Ecosystem Studies. His research interests include multidisciplinary studies of urban ecological systems, the role of disturbance and succession in wild and managed vegetation, and the role of structural boundaries in the function of ecological systems. These interests play out in the Baltimore Ecosystem Study, a long-term ecological research program, and the integration of riparian and upland savanna systems in the Kruger National Park, South Africa.

NICHOLAS A. ROBINSON is the Gilbert and Sarah Kerlin Distinguished Professor in Environmental Law and codirector of the Center for Environmental Legal Studies at Pace University School of Law (New York), where he teaches conservation law. He is currently the legal advisor and chair of the Commission on Environmental Law of the International Union for the Conservation of Nature and Natural Resources. He served as the deputy commissioner and general counsel of the New York State Department of Environmental Conservation from 1983 to 1985. He is author of *Environmental Regulation of Real Property* (1982) and other works. He has a J.D. from Columbia University School of Law (1970) and an A.B. from Brown University (1967).

RICARDO ROZZI is a Chilean ecologist and environmental ethicist whose main project is the integration of philosophy and sciences within the field of conservation biology. He holds an M.S. in ecology from the Universidad de Chile and an M.A. in philosophy from the University of Connecticut.

In 1994, in collaboration with a group of Chilean and foreign ecologists, Mr. Rozzi cofounded the Institute of Ecological Research Chiloe, an academic institute that aims to link ecological knowledge and application through direct interaction with local communities in southern Chile. Since founding the institute, he has served as coordinator of ecological education and the interface with the social community. He has written educational materials for the Chilean Ministry of Education and schools and scientific and philosophical articles in Chilean and international journals. He is working on his Ph.D. dissertation in biological and cultural conservation in the forest region of southern Chile. Mr. Rozzi serves as the contact person for South America in the International Society of Environmental Ethics.

RICHARD W. SAGE, JR., is associate director of the Adirondack Ecological Center (State University of New York College of Environmental Science and Forestry) and chair of the Town of Newcomb, New York, Planning Board. He has 31 years of experience in wildlife and forestry research and management in the Adirondacks. His recent research includes examining winter feeding of deer in the Adirondacks, ecology of Adirondack black bears, wildlife habitat characteristics associated with public and private lands, management and regeneration of northern hardwood forests, and development of population and ecosystem models for deer, bear, and vegetation in the Adirondacks. He is actively involved in working with local government, state and federal agencies, forest industry, and environmental groups throughout the Adirondacks.

JOSEPH L. SAX is the James H. House and Hiram H. Hurd Professor of Environmental Regulation at the University of California, Berkeley. After working for the U.S. Department of Justice and in private practice in Washington, D.C., Professor Sax began teaching law at the University of Colorado in 1962. In 1966, he moved to the University of Michigan, where he became the Philip A. Hart Distinguished University Professor. He joined the University of California faculty in 1986.

In 1994, Professor Sax was appointed to serve in President Clinton's administration as the counselor to the secretary of the interior and deputy assistant secretary for policy at the U.S. Department of the Interior. In this post he has worked on, among other things, the reauthorization of the Endangered Species Act.

Professor Sax has served as a consultant or board member of 19 different environmental public service organizations and was awarded an honorary Doctor of Law degree by the Illinois Institute of Technology. His major publications include *Mountains Without Handrails* (1980), *Water Law: Planning and Policy* (1968), *Water Law: Cases and Commentary* (1965), and *Defending the Environment* (1971).

CHRISTINE L. SCHADLER is a Ph.D. candidate in natural resources and environmental history at the University of New Hampshire, where she is writing a dissertation on environmental impacts of the New Jersey Parkway. For her M.S. at Antioch New England Graduate School, she wrote "Recovery of the Eastern Timber Wolf in Michigan: Biological and Social Perspectives."

RODGER SCHLICKEISEN is president and chief executive officer of Defenders of Wildlife, a national nonprofit environmental organization whose mission is to protect endangered species and conserve biological diversity. He is a member of the Boards of the Natural Resources Council of America of the national League of Conservation Voters (LCV). He chairs the LCV's Political Committee, which works to elect a proenvironment Congress. He also chairs the Partnership Project, a nonprofit corporation established in 1999 to help build a more unified and effective national environmental movement. He holds a bachelor's degree from the University of Washington, a master's from the Harvard Business School, and a doctorate from George Washington University. He was born in Houston, grew up in Washington and Oregon, and resides in Alexandria, Virginia.

VIRGINIA A. SHARPE is deputy director and associate for biomedical and environmental ethics of the Hastings Center. She is also an adjunct faculty member at Vassar College, where she coteaches an interdisciplinary course on earth system science and environmental justice. She has written on numerous bioethical issues, including environmental sustainability, environmental justice, adverse patient outcomes, responsibility for medical error, the meaning of "appropriateness" in patient care, and the fate of reproductive service in hospital mergers. She recently published *Medical Harm: Historical, Conceptual, and Ethical Dimensions of Iatrogenic Illness* (1998). Dr. Sharpe received her bachelor's degree from Smith College and her Ph.D. in philosophy from Georgetown University.

ELI THOMAS is Haudenosaunee (Iroquois 6 Nations), Onandaga Nation (People of the Hills), Wolf Clan from the northeastern woodland. For the last fifteen years, he has been an artist and a visual narrator. He speaks, writes, and narrates his artwork as an artist in residence in New York state schools. On his land, he has a captive arctic timber wolf, two pure timber wolves, and wolf hybrids to promote wolf education.

CLARK WOLF is an associate professor of philosophy and an active member of the Environmental Ethics Certificate Program at the University of Georgia. He is writing a book on intergenerational justice, and his work in environmental ethics has covered issues of property rights, economic development, resource and welfare economics, and human population.

Index